# Recoding Life

This book addresses the unprecedented convergence between the digital and the corporeal in the life sciences and turns to Foucault's biopolitics in order to understand how life is being turned into a technological object. It examines a wide range of bioscientific knowledge practices that allow life to be known through codes that can be shared (copied), owned (claimed, and managed) and optimised (remade through codes based on standard language and biotech engineering visions).

The book's approach is captured in the title, which refers to 'the biopolitical'. The authors argue that through discussions of political theories of sovereignty and related geopolitical conceptions of nature and society, we can understand how crucially important it is that life is constantly unsettling and disrupting the established and familiar ordering of the material world and the related ways of thinking and acting politically. The biopolitical dynamics involved are conceptualised as the 'metacode of life', which refers to the shifting configurations of living materiality and the merging of conventional boundaries between the natural and artificial, the living and non-living. The result is a globalising world in which the need for an alternative has become a core part of its political and legal instability, and the authors identify a number of possible alternative platforms to understand life and the living as framed by the 'metacodes' of life.

This book will appeal to scholars of science and technology studies, as well as scholars of the sociology, philosophy, and anthropology of science, who are seeking to understand social and technical heterogeneity as a characteristic of the life sciences.

**Sakari Tamminen** is an Adjunct Professor of Science and Technology Studies (Anthropology of Science and Technology) at the University of Helsinki, Finland, and co-editor of *Bio-Objects: Life in the 21st Century*.

**Eric Deibel** lectures STS to engineering students at Bilkent University.

# Recoding Life

## Information and the Biopolitical

**Sakari Tamminen and Eric Deibel**

First published 2019
by Routledge
2 Park Square, Milton Park, Abingdon, Oxon OX14 4RN

and by Routledge
711 Third Avenue, New York, NY 10017

*Routledge is an imprint of the Taylor & Francis Group, an informa business*



*British Library Cataloguing-in-Publication Data*
A catalogue record for this book is available from the British Library

*Library of Congress Cataloging-in-Publication Data*
A catalog record for this book has been requested

ISBN: 978-1-138-22557-2 (hbk)
ISBN: 978-0-367-89731-4 (pbk)

Typeset in Times New Roman by
Apex CoVantage, LLC

# Contents

# Tables

# Figures

# Acknowledgements

This book would not have existed without the preparatory work done in the COST Action 'Bio-objects and Their Boundaries', spearheaded by the BioStep Group. The series of meetings over several years was a rare chance to cement relations with a varied group of scholars, who brought perspectives from different nationalities and fields of specialisation. We wish to acknowledge the significance of such a setup – this book is what it is precisely because of the opportunity to focus on working with people rather than thinking primarily in terms of outputs.

## Sakari Tamminen

Families are more than just families. To my biological family, my wife, and my daughters Mai and Ines: apologies for the late nights at the computer and thank you for your patience. Love you. To my other family, the family of intellectuals in and outside of academia: thinking is a collaborative effort, and so is this book too. You know who you are, and I want to thank you for your support along the way in life, in dialogue, in thought. I also want to extend my gratitude to the Gemic fellows J.Suikkanen and O.Utti, all intellectuals working with us, and to all the members of the HK Club. What a great joy in life to have you around.

I also want to thank all the experts, scientists, people who generously found time for interviews and kindly shared their thoughts about biology, informatics, and future. Finally, I wish to thank Academy of Finland for the grant 'Vital Digital' that allowed me to do the research and enable me to write this book with Eric.

## Eric Deibel

To the colleagues, friends, and family who have continued to support me over the years: I hope you know of my gratitude already. Here I wish to focus on one person in particular: my late father. I am very much aware that my intellectual views of the world, including those expressed in this book, are inspired by his life and the example he set. Firstly, my interest in alternatives follows his example: he could have easily passed for a DIY biologist *avant-la-lettre*, if we count his huge aquarium, full of life and technology, which was the soul of our house. It was designed with the vision of an engineer, constantly searching for spare parts and new breeds through constant interaction with his fellow hobbyists. Secondly,

I have come to realise that it was because of him that the second-hand copy of Foucault's *History of Madness* made such an impression on me as a teenager. I grew up with the understanding that his life was a miracle, while his condition was caused by medical malpractice. The example he set was his self-sufficiency while surrounded by the many small indignities he had to suffer. My views of how science and other facets of society blend are heavily influenced by his condition – or, rather, by how the human condition should be the centre of our concerns as scholars. Yet this quickly becomes overtly extravagant when coming into contact with the force field of constant conditioning that is needed to reproduce academic hierarchies. The period wherein this book was conceived and written brought me closer to his memory: how getting better cannot happen without dignity, our own as well as by working hard to lift the constant state of insufficiency and minor violations we all relentlessly impose on others if we are too selfish – on those that are ill and are trying to survive as well as those that can live their lives to try and make something of themselves.

# 1 Recoding life

## Information and the biopolitical

The new Invitrogen™ GeneArt™ CRISPR Search and Design Tool allows you to quickly search our database of >600,000 predesigned CRISPR guide RNAs (gRNAs) targeting human and mouse genes or analyze your sequence of interest for *de novo* gRNA designs using our proprietary algorithms. Up to 25 gRNA sequences per gene are provided with recommendations based on potential off-target effects for each CRISPR sequence. Once you've selected the optimal gRNA designs, you may purchase your gRNAs and other recommended products for genome editing directly from the Web tool.

–www.thermofisher.com/

But as we go into some combination of outer and inner space, ourselves a combination of carbon- and silicon-based life, we still have questions pertaining to the manner in which we will pursue our explorations. Will we be a well-stirred homogenous 'optimal' monoculture, or will we be a cacophonous anarchy of self-experiments – or something in between?

–George Church, in *Regenesis: How Synthetic Biology Will Reinvent Nature and Ourselves*

### Introduction

Deoxyribonucleic acid, or DNA, has been a contested object of exploration, especially since the conclusion in 1953 that it exists in the form of a double helix. It once was seen as the molecule that unlocks secrets of life that have vexed humankind since the dawn of time (Fox Keller 2000). By the same token, however, the discovery of DNA quickly dislocated the idea of 'life' from its historically situated seat within the realm of the transcendental or divine, relocating it with the immanence of corporeal matter embodying informatic patterns and codes (Doyle 2003). While this development has been seen by many as a 'reductionist' one, we claim that the opposite is true: it is only today that the potentiality of life has proliferated and been released from the traditional forms and functions reserved to it within natural history and the life sciences.

Life is today escaping its confinement to the cell, the molecule, or the idea of the natural code and preferred patterns of expression shaped by evolutionary forces, for it has been relocated within a number of *new* bodies. Life is found in new corporeal species bodies reconfigured through information and code, in the social bodies of knowing and manipulating innovated life forms, and in the institutional bodies of governance reshaping the ways of composing our post-modern communities and 'proper' ways of living together. Together these are expressing the seemingly bottomless generative potentiality hidden in the idea of life – today expressed increasingly through the language of utility and exchange value.

As Thermo Fisher – a behemoth life sciences instrument company with annual turnover above 17 billion dollars and more than 50,000 personnel on staff, responsible for, as an example, the Invitrogen genome analysis and editing tool and its marketing – suggests, searching for various genes and editing the code of life has become as easy as 'surfing the Web' and purchasing products on any given Web shop found online. The redesign of genes, perhaps of whole genomes in the near future, is made available for anyone with the interest and an accepted credit card. Below Thermo Fisher's statement at the beginning of the chapter, we quoted a claim by George Church, one of the most prominent geneticists *cum* molecular engineers and a professor at Harvard and MIT fellow. What it suggests is that we must soon take a stance on how we, *Homo sapiens*, want to see ourselves as a species in the future. Are we going to edit genes or rebuild our makeup in a coordinated effort to fashion a superior form of life immune to diseases and ageing? Or are we going to transcend our species boundaries and let self-experiments on genetic modification take over? Whichever path we choose, he suggests, the life of Earth is profoundly changing via genetic engineering or synthetic biology, and we are witnessing, as a result, nothing less than a total *regenesis* of life.[1]

There are constant reminders of the steps being taken towards this kind of regenesis, or at least the editing of the human genome. One of these is the UK Human Fertilisation and Embryology Authority (HFEA), which in early 2016 granted permission to edit genes in human embryos for research purposes to a team of UK scientists at the Francis Crick Institute, led by Dr Kathy Niakan. In February 2016 *Nature* reported as follows:

> The HFEA has approved an application by developmental biologist Kathy Niakan, at the Francis Crick Institute in London, to use the genome-editing technique CRISPR – Cas9 in healthy human embryos. Niakan's team is interested in early development, and it plans to alter genes that are active in the first few days after fertilization. The researchers will stop the experiments after seven days, after which the embryos will be destroyed. The genetic modifications could help researchers to develop treatments for infertility, but will not themselves form the basis of a therapy.[2]

The editing of human genetic material had been under moratorium for several years, as no overarching consensus on ethical or philosophical principles has been found among biotechnologically advanced countries. Now, however, the UK authorities have decided that genetic editing for research purposes is ethically defensible. In the same article, the Crick Institute's director, Paul Nurse, explained the value of the gene editing: 'Dr Niakan's proposed research is important for understanding how a healthy human embryo develops and will enhance our understanding of IVF success rates, by looking at the very earliest stage of human development'. While embryo editing is now allowed only for research purposes, it is not difficult to imagine these techniques going beyond the laboratory or medicine, entering clinical trials or other fields. This is to be expected if only because techniques such as gene editing are not regulated in many countries. Not long after the *Science* piece, in July 2017, embryo gene editing hit the news in the United States, as the Oregon Health & Science University reported it had successfully edited embryo genes (MIT Technology Review 2017). One day later, the *New York Times* reported how other scientists commented on this experiment and directly cited one prominent genetics expert, Dr. Eric Topol, director of the Scripps Translational Science Institute in La Jolla, California, who said that gene editing of embryos is 'an unstoppable, inevitable science, and this is more proof it can be done' (*New York Times* 2017). This makes the question posed by Church all the more pertinent. Will it be a rogue scientist group who claim to be making the first 'perfect human baby'? Will we soon see DIY biologists engineer biopunk life forms in their garages? What will such life forms look like? And, perhaps, how should we as social collectives react to this? (See, for example, Wohlsen 2011; Delfanti 2013.)

The recent developments in genetic engineering tools, knowledge, and their use in improving species of life are only the logical continuation of the historical trajectory of biology and biosciences. Gradually the question about finding the 'truth' about life has shifted toward a question about the utility of such techniques in making life forms perform a particular task in a more efficient, optimised way. This is a result of 'the century of the gene' (Fox Keller 2000), brought to us via new innovations for the manipulation of life both intra-cellular and extra-cellular and through particular biopolitics reframed around the optimisation of not only the individual and the species but the way in which individuals and species develop, perform, and regulate themselves (or not). Such biopolitics targeting life *not as it is but how it should be* is a question of interests, or, rather, it is a politics of interests that fundamentally defines how the 'code of life' will be compiled and executed at intra- and extra-cellular levels across species and taxa (see Foucault 2002).

This book examines how genetic engineering, synthetic biology, and different life forms being generated constitute the emerging forms of life of the early twenty-first century's global culture (Fischer 2003). Here, 'life', and truth about life, is increasingly verified within a field of competing interests where their value and utility is premised on the fading of species bodies and their borders. Accordingly, bodies of knowing, their techniques, their governance, and the very ways of speaking about life and its representation are reshaping the capacity for optimisation and their exchange value, many times called the sphere of the neoliberal 'bio-economy' or, occasionally, biocapitalism (see Sunder Rajan 2006; Cooper 2008; Helmreich 2008). Life in the form of informational media (Thacker 2003) should be understood as a newly framed matter of concern (Latour 2004) that arises from how bio-objects are being assigned vital functions and powers holding utility value and, therefore, potential value in exchange (Vermeulen et al. 2012).

All this is true, yet the framing and regulation of such emergent forms of life is always already a subject of a global politics of interest, wherein power is negotiated through particular configurations entrenched in the deep will to explore the limits of life's value. This signals a regenesis not only of life but of a global biopolitics co-constituted through the *metacodes of life*: revealed through its discourses, techniques, circuits, and their intersections alongside the interactions that support particular types of forms of research, exchange, and governance.

## Exploring metacodes

The idiom of exploration is omnipresent throughout the life sciences, and there is always a multiplicity of objectives to accomplish and at stake. Crucially, the question of life – molecular, genetic, synthetic, and beyond – is fundamentally unstable. An example of this is how these objectives are imagined by the authors of the breakthrough report on the completion of the map of the human genome. In their celebratory article announcing the complete map, the International Human Genome Sequencing Consortium on Initial Sequencing and Analysis of the Human Genome quoted T.S. Eliot's *Four Quartets* (15 February 2001, *Nature* 409). Looking beyond their publication on the human genome, they concluded:

> Finally, it has not escaped our notice that the more we learn about the human genome, the more there is to explore: We shall not cease from exploration and the end of all our exploring will be to arrive where we started. And know the place for the first time.

As expected, the initial completion of the map did not end the exploration. Instead, the decade and a half that has passed since has shown an exponential growth in mappings: numerous species of mammal, plant, bacterium, and so forth or microbial genomes containing billions of base pairs have been charted and mapped into databases. Moreover, the practise of mapping as a means of generating 'basic knowledge' has turned into a practise of modelling the increasingly complex behaviour of genes, proteins, and cells. Laboratories and infrastructures around the world are expanding, facilitating the exchange of information, instantaneously, for diverse purposes, in what is best characterised as an industry informed by biological science and seeking to re-materialise the information through genetic engineering.

Therefore, it was certainly appropriate that the authors of the report claimed that there is always more to explore, but what they suggest is that the end is 'to arrive where we started'. This refers to the subject of the exploration as life itself: its inner workings mapped and revealed by science imagined as a domain with a beginning that always already was an end in itself. In this sense, the advance of the life sciences moves from an ahistorical past towards a horizon that is constantly being reached, moving towards the realisation of the promise of a future in which technique subdues life and nature. Such an exploration goes onwards (there is always more to explore) as if it is only a matter of time before the furthest extent of knowledge arrives where it started: it is premised on a transhistorical understanding that is a crucial feature of what we call the *metacode of life*. In a sentence, the metacode of life refers to the centrality of DNA in informatic formats as the condition of the contemporary politics of how the biological informs the digital and *vice versa*.

Maps are powerful techniques for representing worlds or spatiotemporal configurations of specific relationships between human and other beings that are not created out of nothing or only life itself. Maps are exemplary, as already observed by Donna Haraway when she stated that 'maps are models of worlds crafted through and for specific practices of intervening and ways of life' (Haraway 1997, 135). In this sense, the exploration imagined by the authors of the report on the human genome project – whose aim 'to arrive where we started' – suggests that once 'life's inner workings are mapped and revealed by science', biology as the *logos* of *bio* will have been transformed, questioned, re-articulated once again. We will mark such transhistorical understandings as *metacodes of life*, as a starting point and structure from which to rethink DNA as information and embodied forms as they are explored and mapped in ways that mediate spaces, relationships, and actors through time.

The citation that marked the completion of the map of the human genome is particularly telling, as its lines taken from T.S. Eliot's *Four Quartets* refer to Dante and his story about the emissary of the living who was guided by Virgil and Beatrice among the dead. The passage cited underscores the continuous nature of exploration and refers to a mystical interconnection of the past, present, and future. Such a perspective on the first completed map of the human genome implies an exploration that similarly breaks free from the boundaries of previous journeys. In this case, it is an exploration of 'the human' and its 'limited-edition formats' such as the 'body'. The analogy changes the kind of exploration and what is at stake. It includes Inferno, Purgatory, and Paradise rather than an easy choice between heaven and hell or between salvation through the inventions of the life sciences and damnation through genetic engineering becoming like Frankenstein: 'a thing such as even Dante could not have conceived' (Shelley [1818] 2006, 61).

We are told, again and again, about how new technologies will deliver us more food, health, and wealth and improve the soil, even the weather. Soon everyone will win, when the life sciences realise their promise as drivers of knowledge economies – as a global bio-economy; a knowledge-based bio-economy; a bio-based economy; or similar arrangements wherein ecology, economy, and biology find each other. Following these guides means that we are presented with a journey that is about genetic exploration that has as its end a sustainable future that is reached by realising the long list of innovations in every imaginable area that will guarantee employment and sustainability without compromising our agricultural productivity, global food security, focus on renewable energy, and potential solutions to climate change (see Chapter 7).

Sometimes this type of 'end' to the exploration is presented as a myth of origins: a return to the mists of time when everything was always already 'bio-based' rather than the fossil resources that power today's industrialised civilisation. This is a painting of a lost Arcadia wherein only biological resources were used for food, shelter, transportation, and so on (see Veraart et al. 2011). Mostly, however, historical continuity needs to be re-established in the present, which invariably means that the commercialisation of the biological sciences is a premise for the extensive wish list attached to the usage of sustainable and renewable biological resources.

On one hand, the 'end' of the exploration is taking on the shape of a comprehensive alternative, with the life sciences as the driver of a systemic transition to a world wherein societies are revitalised after suffering economic crises, along with their citizens, who are in harmony with themselves, each other, and natural environments. On the other hand, the counter-narratives mirror the increasing magnitude of the story, as found in many of the policy documents on the life sciences and the knowledge economy. Instead of harmony and the unity of science, society, and nature, this is a discourse of conflict and competition looming in the near future, when resources (such as water and land) will be at stake, new sectors will displace others (as with fuel and food), and the fortune of nations might need to be guaranteed by various means as entire continents surge ahead or fall behind (discussed in Chapter 7). Underneath the heavenly vision of new biological, economic, and ecological harmony we find a hell that is populated with the same characters that are held responsible for the loss of nature's integrity: new monocrops, big pharma, big data, and the plundering of biological wealth.

While two diametrically opposed conclusions are being drawn in looking at the biological future, both are narrated as a sweeping and dramatic epic with geopolitical consequences. In this sense, they are perfectly symmetrical in their focus on the need to overcome planetary constraints and in their agreement about how humans will have to act as a species, and at a scale that matches that of the numerous global crises of the present. At the centre of both stories we find a reverence for the desire and drive to overcome whatever limits there are to life as a technological creation and to redesign species bodies at will. After all, each appeals to a 'return to nature', which assumes that the journey has an end that, whether good or bad, is premised on a logical coherency that is simply not there, neither in the knowledge about DNA nor in the political philosophies about the natural world and human nature (see Hayles 1999; Žižek 1999; cf. Fukuyama 2002; Habermas 2003).

What could be more appropriate for the contemporary life sciences than how an emissary of the living is guided through the rings of hell, purgatory, and heaven? Certainly, the guides are different, but even without Dante and his guides, the end to the exploration is still a cosmological order wherein everything is assigned the right place – this time being reconfigured biologically, genetically, or by informatic patterns encompassing all living beings. Once again we get to bear witness to a renewal of practises of identifying, observing, classifying, ordering, representing, analysing, redoing, and rerouting life and death, the living and the non-living. This time we follow guides who suggest that there is a new era of techniques of life, thereby establishing a type of continuity that spans from Dante's vision to today, passing by Diderot's *Encyclopédie* and more recently Wikipedia and the human genome as a repository of knowledge of what humans are. Each in its own way has been an effort to incorporate all of the knowledge scattered over the surface of the earth, showing a continuity in technique, skill, tools, and *techné* – each pointing to technology as an enduring part of what it means to be *Homo sapiens*, as individuals and as a species.

As an 'end' to the exploration we will seek to rethink the return to nature as a myth of origins that enables us to recontextualise how bio-objects are escaping the natural history of species. Such an exploration of biological entities (and knowledge of such entities) does not shy away from notoriety – quite the opposite. We seek to affirm how controversy ends up being the norm with new techniques of life. Simultaneously, we are genuinely interested in the mapping, modelling, and programming practises that are characteristic of the rapid transformation of the contemporary life sciences. Ultimately, it is in this reconfiguration of life and bodies that the contemporary cosmological order is performed, as a scientific practise and as a biopolitics that draws its strength from its intimate relationship to the desire 'to optimize, enhance and renormalize what counts as biological' (Thacker 2003, 76).

## Mapping information and the biopolitical

To present genome mapping as the core of an exploration of historical significance establishes a historical comparison to the age of exploration: to the time of world maps that had gaps in the middle of familiar shorelines (see Zwart 2009).

These maps are more familiar to most of us, as they occupy a large part of colourful picture-filled books on the history of mapmaking, which show, page by page, how natural harbours and inland rivers become visible, new spaces yielding to explorers and missionaries and thereby revealing a world with unknown territories. Someday, the final pages of such photo books might depict the human genome or other species' genomes. And why not? The latter deliver strange fruits and artefacts as well as tales of riches, miracles, and even exotic peoples whose very nature has been captured in beautiful photographs. Yet such metaphors, of exploration and mapmaking, are misleading when presented in isolation from how the language wherein biological life was traditionally understood was already becoming informationalised in the mid-twentieth century (see Kay 2000; Fox Keller 2002; O'Malley and Dupré 2007; Mackenzie 2005; Doyle 1997). Such a historical analogy prioritises the pervasive language of life wherein DNA is code, to be programmed as if life were the equivalent of information that can be stored, transmitted, and shared.

With the application of database technologies to the biological record, prioritising the creation and usage of information resources and technologies has as its result a biological record that takes the database as its organising principle. As Geoffrey Bowker put it, biology and information each bootstrap the other (Bowker 2005). This is our perspective as well; a different type of reading of the history of biology takes precedence: one based on an underlying politics of science and technology that takes its shape at the intersection of two fields built on shifting unstable configurations of life, facts, data, and artefacts that cut across conventional boundaries between natural and artificial, living and non-living (see Vermeulen et al. 2012). It is in this register that we approach the intersection of life and information in terms of 'the biopolitical'. This term from the subtitle of the book refers to Michel Foucault and his explanation of the concept in the last chapter of *Society Must Be Defended* (Foucault 2003, 240–263).

The concept does not refer only to politics of science and technology that happens to be about biology, imagined as a more or less singular frontier of knowledge or moment of progress at which nature is overcome at last. Instead, our understanding refers to how 'biopolitics deals with a problem that is at once scientific and political, as a biological problem and as power's problem' (ibid., 245).

Foucault's concern with biology is an element of his lectures on the transformation of the power over and rights of alive and dead. The point is, therefore, not that power and the biological sciences are indistinguishable; instead, 'power's hold over life' is understood by distinguishing between, on one hand, how sovereignty emerged from early-modern traditions of political theory and, on the other, the emergence of techniques of power centred on the body in the nineteenth and twentieth centuries. This is crucial; what it does is bring to the foreground that the emergence of new techniques of power is inescapably tied to processes of permeation and saturation whereby the relationship of sovereign rule with the multiplicity of men is dissolved to the level of individual bodies that can be kept under surveillance, trained, used, and – if need be – punished (ibid.).

Consider again the age of exploration and the analogy with the mapping of genomes. This time, we take a different beginning, starting with how 'collecting' received 'the status of a science' (Parry 2004, 20–22). As Bronwyn Parry explains in *Trading the Genome*, within a specific cultural moment, the study, classification, and categorisation of samples, specimens, and artefacts became more than the 'culture of collecting' it had been before. Natural history was shaped as a scientific field through successive search missions to refine its focus and scope. The constant redefinition was possible as

> each voyage brought back a systematically organized body of information about the coastlines, flowers, fauna, language, and cultures of distant peoples. This information could be employed to recreate within particular dedicated spaces in Europe – such as museums and laboratories – a scaled-down version of the world that could be surveyed panoptically.
>
> (ibid., 20–22)

These voyages returned with knowledge about events and places elsewhere as well as with the materials that were to be concentrated in collections. Of course, not all materials were brought back in their original form; this was neither possible nor useful. A coastline, for example, was better translated into an artificial form such as a map's surface. Similarly, there were botanical illustrations, stuffed animals, and tropical plants that could not be transported but could be described and categorised in a way that could effectively 'stand in for' or 'represent' the organism in question (Parry 2004, 23). Parry explains how the capacity to transfer materials from the centre to the periphery was the function of many different technologies, 'bringing natural entities – or at least some of their key properties – home' (ibid., 23). For us, such an interpretation of the metaphor of mapmaking implies that the biopolitical is seen in terms of a different kind of historical continuity. The point is no longer that there was a golden age of exploration or a neocolonial nightmare; instead, the historical analogy revolves around configuration of techniques, knowledge, and life forms that are performed in specific settings and yet can be seen as global in scope.

More specifically, Foucault's analysis envisaged power as exercised through interventions in the rate of birth, longevity, sanity, illness, reproduction, and so forth. It was only late in his life that Foucault started using the term 'biopolitics'. It was introduced in the lectures dealing with biopolitics, found in *Society Must Be Defended* (1975/6) and *The Birth of Biopolitics* (1978/9), wherein he carefully distinguished his interpretation from conventional understandings of sovereignty in terms of the actions of nations as a foundational category or myth that can or should be identified with stable economic, political, and public interests. It is this relation to political theory that remains key, especially when one is concerned with emerging technologies. Not only do conventional notions of territorial sovereignty remain powerful in their relations to the life sciences (see Chapters 4 and 5); the same applies to the technologies of power that emerged later. Also, these need to be understood in terms of how 'pre-critical naïveté holds undivided rule' (see Foucault 2002, 349–351, 2008). Understanding and demonstrating this type of naïveté is crucial since it permeates Foucault's studies of power. It is a constant part of his discussions of the kind of power that was exercised in the nineteenth century through regulatory mechanisms tying the individual to the nation's economy through statistical techniques, forecasts, or estimates or through overall measures that absorb geographical and environmental conditions into the well-being of the population.

We believe that this type of political theory and theorisation is invaluable when one invokes the biopolitical as a perspective on the practises and epistemologies of the life sciences. It is, therefore, by developing such a perspective that this book speaks to the biopolitical and the authors seek to think across diverse fields of bioscientific knowledge practices. We illustrate the theoretical perspective with a wide variety of cases that each paint a biopolitical terrain wherein power is exercised in line with informatic ways of thinking about life and nature. Engaging closely with the naïveté arising out of early-modern thinking about sovereignty, social contracts, and international order, we are able to approach the rapid changes in the life sciences methodologically as well as critically. Indeed, we will show that it is in terms of early-modern naïveté that we can best understand the biopolitical in its relations to expectations surrounding how life is not only being known or decoded but also remade (or recoded) and re-materialised through bioinformatics and material biotechnologies. This all happens within the context of particular metacodes – empirical positivities acting as guiding principles, meanings, values, and practises – that frame what life is, can be, and is made to be through specific epistemic platforms they form today and in the future.

## Life, code, work

> I would like to assure you that, in spite of everything, I really did intend to talk about biopolitics, and then, things being what they are, I have ended up talking at length, and maybe for too long, about neo-liberalism.
>
> –Michel Foucault, *The Birth of Biopolitics* (1978, 185)

As Foucault puts it, talking about biopolitics should not be exclusively tied to neoliberalism. By this we mean that focusing excessively on the status of the sciences within the context of 'bio-capitalism' does not get us very far in terms of theoretical or practical analysis. This would be a limitation, irrespective of whether this recontextualisation (attaching 'bio' to 'capitalism') would announce a 'new epoch' or diagnose a 'symptom' or merely would invoke the basic concepts of political economy, among them value, markets, and the commodity (Sunder Rajan 2006).

We maintain that examining the biopolitical implies that there is no necessary choice to be made between a critique of neoliberalism and studying the life sciences. As Kaushik Sunder Rajan explains, Foucault and Marx are comparable because the latter examines political economy as 'a foundational epistemology that allows us the very possibility of thinking about such a system as a system of valuation' (ibid., 7–13). This is also our position: the possibility of thinking in terms of systems of valuation implies studying facts, data, and objects that 'act' (bio-)politically and that perform socio-technical arrangements, wherein fluidity and perpetual motion undercut established materialities and socio-political orderings. Sometimes this includes situating the life sciences in the context of familiar patterns of financial capital, trade, and industrial production (see Birch and Tyfield 2013). Other times, however, the (im-)materiality of science and technology undercuts political forms of action, organisation, and thinking; leads to credibility being lost or gained (Gieryn 1983); and is implicated in how relationships between science and society are constantly (re-)generated (Latour 1993) and in established ways of thinking and acting politically that can be blocked, reinterpreted, and transformed.

In other words, we maintain that no choice has to be made of whether to think of biopolitics in terms of life or in terms of capital. For example, Stefan Helmreich observes that biopolitics needs to be adjusted because of how the life sciences 'enable different biopolitical constellations, ones not so neatly organized around genealogy and birth, or for that matter through human bodies' (Helmreich 2009, 101). Yet in related types of analysis, the focus shifts back to neoliberalism. A key example is Melinda Cooper's *Life as Surplus*, which gives priority to how strategies of neoliberal biopolitics mirror the life sciences in embracing complexity theories, thereby finding new ways to reduce the 'extraeconomic' to exchange value (Cooper 2008, 10). Both positions, however, are easily criticised in terms of how tenuous the hold of the transformations in the life sciences is over global capitalism. Only to a limited extent can the life sciences' position with regard to global economic and geopolitical arrangements of neoliberalism explain the shaping of bioeconomies (Birch and Tyfield 2013).

The important point, to us, is that the very notion and need for priority setting in 'bio-capitalism talk' – capital versus biocapital, science versus markets, technology versus society, and so on – sits uneasily within a Foucauldian approach. Consider how Foucault examined the formation of fields of modern knowledge in *The Order of Things*. He studied three histories in parallel: how wealth became economics, how natural history became biology, and how the study of language became linguistics. These pieces of epistemological terrain are shown to be mutually constitutive, which is to say that the science of life is misunderstood when represented by singling out a particular history or privileging one of the constitutive parts over the others. Taken together, they demonstrate modernity as a space of representation that is grounded in the 'rigorous and general epistemic re-arrangement' of the nineteenth century (Foucault 2002, 180–182).

This is also the basis for his easily misunderstood claim that there could be no 'life' in the modern biological sense of the term before this time. Famously he argued that 'it is not so long ago – when the world, its order and human beings existed, but man did not' (ibid., 349–351). Such life and such a human were inconceivable in classical science; it is only through the modern representation of life, language, and labour that a subject comes into existence that he calls the 'emperico-transcendental doublet' (ibid.). What is relevant about this view is that it reveals how contemporary debates are premised on a defence or a celebration of how an original or logically coherent conception of the human subject is about to be displaced – to be lost or saved. It offers us an alternative to the way in which a choice between those two viewpoints ends up reaffirming that there is a posthumanity or a post-nature that is soon to arrive as a consequence of the radically changing character of biological modification.

We remain interested in the very ability to comprehend 'life' and 'economy' in their modernist guises, as shaped by particular epistemologies that simultaneously are enabled by and, in turn, enable 'particular forms of institutional structures' (Sunder Rajan 2006, 14). This includes the third example of a modern field of knowledge and the one that Foucault found most important, linguistics. Sunder Rajan refers to 'a shifting grammar of life, towards a future tense' (ibid., 14). Such an understanding of the metacode of life is linguistic in the ability to calculate, showing how different grammars are at play in how living materials are exchanged and circulated as a series of coding operations wherein DNA gets transcribed into RNA.

Taken together, the three fields are each mobilised when digital practices and mapping metaphors give way to genetic engineering and the re-materialisation of information as food, medicine, energy, and other forms of living. Alongside its biology and its economics, this is a process that should be understood as profoundly linguistic in the sense that coding and programming are integral to processes of gene expression and regulation and to sequences that can be cut, spliced, and transcribed in various ways. Ultimately, this is a language that 'not only contains digital instructions that make us who we are' but is written down in ways that are strategically and selectively related to a wide range of 'institutions, procedures, instruments, practises and forms of capitalization', as Nicolas Rose put it (2001, 13–15).

## The structure of the book

The biopolitical horizon opened in this book is no longer the one of nineteenth-century disciplinary techniques nor a calculative biopower enabled by new life sciences. The new biopolitical reality is still in formation. Hence, the outline of this book echoes Gilles Deleuze's closing observation on Foucault's work some thirty years ago, wherein he finished by asking about the new forces beginning to take shape:

> Biology had to take a leap into molecular biology, or dispersed life regroup in the code. Dispersed work had to regroup in third generation machines, cybernetics, and information technology. What would be the forces in play, with which the forces within man would then enter into a relation?
>
> (Deleuze 1999, 109)

While the new biopolitical reality continues to refer to life (new forms of biosciences), to language (code), and to labour (dispersed work) – a triptych that echoes the epistemic fields explored by Foucault in his *The Order of Things* (Foucault [1966] 2003) – simultaneously there are new relations of force, acting on one another and through life, transforming what it means to be living today (see Dillon and Lobo-Guerrero 2009).

Charting out some of the first contours of this new reality is the goal for the next chapter (Chapter 2), by which we seek to turn our position into an approach, detailing our rethinking of the biopolitical, theoretically, conceptually, and methodologically framing the empirical analyses that follow. Starting with what Foucault called 'the problem of sovereignty' in his later work, we retrace his steps to his early work and return to the question about the 'modern space of representation' and its epistemic underpinnings. Drawing from Foucault's early work, we suggest a perspective on the biopolitical extended to the variety of ways wherein power is exercised through the shared and collective social bodies that enter into new relations in consequence of the bioinformatic reconfiguration of humans and non-human life forms.

Chapter 3 looks at the standardisation of biosciences through global digital infrastructures, categories, and methods of representation. The chapter provides background for the idea of a standard biobank language and points out how the dream of a standard biological language has always been one of the aims in the biosciences. Empirically, the chapter analyses the standardisation of the fundamental information model for the first EU-wide research platform, the European Biobanking and Biomolecular Resources Research Infrastructure (BBMRI-ERIC; work led by the Swedish Karolinska Institutet). In the chapter, we claim that the standardisation of a universal biobank language conceptually transforms also the objects of scientific enquiry – human biobanks (what they are) and the objects they hold (the concepts of 'donor', 'sample', and related 'data') – for purposes of developing an information model that is suitable for large-scale collaboration.

Chapter 4 examines the global politics of access, dealing with species of plants as a subject of international agreements and governance mechanisms. Rather than considering 'access' as a push back to the rules of ownership and the commodification of life, the chapter shows how a new generation of 'miracle crops' is being launched as integral to an organisational model wherein global targets for environment-related policy making cannot be dissociated from the aspirations of the life sciences and sophisticated techniques. Not only is the model ineffective, but also thinking across the types of 'access' that are possible shows that exclusivity should be rethought in terms of high-tech forms of gifting, charity, and altruism that are becoming a global norm. The theoretical aim behind the chapter is to show how such a norm expresses a state of exception wherein life and law collapse amidst the constant need to guarantee access within the confines of the complex interplay of exclusive claims – of states over genetic wealth in their territory through intellectual-property protections, state-based regulation, and new governance mechanisms that dissolve earlier distinctions between public and private.

Chapter 5 continues the analyses of how the life and value of living beings has been reframed in recent decades in global political conventions. As were plants, animals used to be valuable as far as they could produce or provide physical goods or labour to humans. Today all living beings are seen as fleshly 'function libraries' for valuable jobs (immunity, chemical-secretion, greater productivity, etc.), easily accessed via bio-informatics and to be appropriated and used for new organisms and purposes. These ideas are codified in the widely recognised Convention on Biological Diversity and Nagoya Protocol, signed by most countries in the world. Together, these global conventions have created a new and globally binding regime for 'labour contracts' affecting all (non-human) living beings. The chapter again foregrounds how today's nature and non-human life are reduced to their coding components, such as 'genetic resources' to be globally indexed, stored, and ultimately put to work for the desired ends. In addition, however, it pinpoints how territorial sovereignty is performed in the context of these powerful global forces that are at play in transforming the materiality and information of genetic resources into objects of a new global nature politics.

The discussion of access continues with Chapter 6's examination of synthetic biology as the most influential extension of the idea and norms of 'Open Source' to the life sciences. The suggestion is that access and openness are needed to encourage collaboration, the sharing of knowledge, and further development of platform technologies. Yet this alternative mirrors a normative space situated at the intersection of informatics and the commodification of life. Digital platforms and genetic techniques underdevelopment are closely related to business models applied for delivering whatever synthetic compound might be in demand on the world market – plastics, chemicals, oil, and the like. After discussing what openness means in its relation to the design community associated with the BioBricks Foundation (BBF), the chapter turns to re-imagining what a rigorous application of Open Source principles to genetic engineering might look like. To that end, we compare the 'minimal-genome' and 'minimal-cell' projects. Their differences enable us to go beyond rhetoric and into the underlying scientific practises to re-imagine whether an Open Source approach has potential to offer an alternative that extends beyond the desire to solve global problems through genetic engineering imbued with conventional ideas of property, ownership, and access to markets and resources.

Chapter 7 engages even more directly with the life sciences as an alternative space of representation for key concepts of the new bioepistemic reality. First, the 'global bio-economy' is examined as a policy language and framework that is most powerfully advocating the life sciences as the drivers of a transition to major changes in society – that is, a transition to a world of sustainable societies. Invariably, its many goals rely on the ability to engineer malleable biomass into a sustainable and renewable source of energy, food, and materials (e.g., bio-plastics), all at the will of the designer. Crucially, however, such a vision of the future is not solely an extrapolation from the promises of synthetic biology as to the global targets that it is employed to contribute to reaching; the visionary work is increasingly a reflection of an alliance of chemical companies and synthetic biology start-ups seeking to industrialise specific commodity chains that are to remain exclusive. It is within this setting that the chapter continues by comparing and contrasting 'biohacking' and 'Open Source seeds'. The former is an effort to transform the life sciences by opening up the laboratory and making experimentation inclusive, while the latter approach represents seeking to remove the restrictions that are imposed on the usage of plants in agriculture, extending open licencing for seeds. Accordingly, the Open Source Seed Initiative (OSSI) does not seek to justify the existing restrictions or to mediate them by sharing knowledge and exchanging information; rather, they actively strive to develop new strategies to roll back the commodification of seeds. While this is difficult to accomplish, OSSI demonstrates that a counter-economy is entirely practical in one of the most contested life sciences domains, suggesting that a similar approach might be applicable to living and working with genetic materials in other formats and bioeconomies.

The concluding chapter returns to the notion of metacodes. As is noted earlier, this concept refers to the question of new forces shaping the reality of our bio-infused present and the forces we as *anthropos* are entering into relationship with. We suggest that these forces form new epistemic underpinnings that at once present a new challenge for understanding our reality and give new tools for critique and social change. Finally, they constitute fragments of an experimental theoretical discourse, a way of testing how well new politics, forms of living, and ways of governing work together. We summarise how our work fits within the wider academic context of popular theories today that have addressed the rise of the bio-economy, biocapital, and bioproperty (e.g., Sunder Rajan 2006; Thacker 2005; Cooper 2008). Our argument, woven throughout the book, is that these do not fully account for what is happening to life today. We need to return to 'dead theories' (e.g., ideas of Rousseau, Hobbes, and Marx) out of which social science emerged to go beyond critique and thereby imagine a future that is grounded in the diversity of politics of nature as unfolding today, alongside an understanding of how the forces we are entering into relationship with are not so new after all.

## Notes

1 Church's vision is a rephrasing of Marshall McLuhan's vision from the mid-1960s of the effects of new media technologies in enhancing our physiological and perceptual capacities, now in connection with the reengineered human body. While McLuhan claimed that

'in the electric age we all wear all mankind as our skin' (McLuhan 1964, 47), Church envisions the capacities of a biologically altered species. This perspective is elaborated upon fully in Chapter 3.

2 See www.nature.com/news/uk-scientists-gain-licence-to-edit-genes-in-human-embryos-1.19270 (accessed on 1.12.2016).

## References


Birch, K. and Tyfield, D. (2013) Theorizing the Bioeconomy: Biovalue, Biocapital, Bioeconomics or . . . What? *Science, Technology, & Human Values*, 38(3): 299–327.

Bowker, G. C. (2005) *Memory Practices in the Sciences*. Cambridge, MA: MIT Press.

Church, G. and Regis, E. (2012) *Regenesis: How Synthetic Biology Will Reinvent Nature and Ourselves*. New York, NY: Basic Books.

Cooper, M. (2008) *Life as Surplus: Biotechnology and Capitalism in the Neoliberal Era*. Seattle, WA: University of Washington Press.

Deleuze, G. (1999) *Foucault*. London: Continuum.

Delfanti, A. (2013) *Biohackers: The Politics of Open Science*. London: Pluto Press.

Dillon, M. and Lobo-Guerrero, L. (2009) The Biopolitical Imaginary of Species-Being. *Theory, Culture & Society*, 26(1): 1–23.

Doyle, R. (1997) *On Beyond Living: Rhetorical Transformations of the Life Sciences*. Stanford, CA: Stanford University Press.

Doyle, R. (2003) *Wetwares: Experiments in Postvital Living*. Minneapolis, MN: University of Minnesota Press.

Fischer, M.J. (2003) *Emergent Forms of Life and the Anthropological Voice*. Durham, NC: Duke University Press.

Foucault, M. (2002) *The Order of Things: An Archaeology of the Human Sciences*. London: Routledge.

Foucault, M. (2003) *Society Must Be Defended*. New York, NY: Picador.

Foucault, M. (2008) *The Birth of Biopolitics*. New York, NY: Palgrave Macmillan.

Fox Keller, E. (2000) *The Century of the Gene*. Cambridge, MA: Harvard University Press.

Fox Keller, E. (2002) *Making Sense of Life: Explaining Biological Development With Models, Metaphor and Machines*. Cambridge, MA: Harvard University Press.

Fukuyama, F. (2002) *Our Posthuman Future: Consequences of the Biotechnology Revolution*. New York, NY: Picador.

Gieryn, T.F. (1983) Boundary-Work and the Demarcation of Science From Non-science: Strains and Interests in Professional Ideologies of Scientists. *American Sociological Review*, 48(6): 781–795.

Habermas, J. (2003) *The Future of Human Nature*. Cambridge: Polity.

Haraway, D.J. (1997) *Modest_Witness@Second_Millennium.FemaleMan©_Meets_OncomouseTM: Feminism and Technoscience*. New York, NY: Routledge.

Hayles, N.K. (1999) *How We Became Posthuman: Virtual Bodies in Cybernetics, Literature, and Informatics*. Chicago: University of Chicago Press.

Helmreich, S. (2008) Species of Biocapital. *Science as Culture*, 17(4): 463–478.

Kay, L.E. (2000) *Who Wrote the Book of Life? A History of the Genetic Code*. Stanford, CA: Stanford University Press.

Lander, E.S., Linton, L.M., et al. (2001) Initial Sequencing and Analysis of the Human Genome. *Nature*, 409(6822): 860–921.

Latour, B. (1993) *We Have Never Been Modern*. New York, NY: Harvester Wheatsheaf.

Latour, B. (2004) *The Politics of Nature: How to Bring the Sciences Into Democracy*. Cambridge, MA: Harvard University Press.

Mackenzie, A. (2005) The Performativity of Code: Software and Cultures of Circulation. *Theory, Culture & Society*, 22(1): 71–92.
MIT Technology Review. (2017) *Rewriting Life*. First Human Embryos Edited in U.S. Available at: www.technologyreview.com/s/608350/first-human-embryos-edited-in-us/ (accessed on 11.8.2017).
*New York Times*. (2017) In U.S. First, Scientists Edit Genes of Human Embryos, 27 July 2017. Available at: www.nytimes.com/aponline/2017/07/27/health/ap-us-med-embryos-gene-editing.html?_r=0 (accessed on 11.8.2017).
O'Malley, M. and Dupré, J. (2007) Size Doesn't Matter: Towards a More Inclusive Philosophy of Biology. *Biology and Philosophy*, 22(2): 155–191.
Parry, B. (2004) *Trading the Genome: Investigating the Commodification of Bio-information*. New York, NY: Colombia University Press.
Rose, N. (2001) The Politics of Life Itself. *Theory, Culture & Society*, 18(6): 1–30.
Shelley, M. (2006) *Frankenstein; or, the Modern Prometheus*. London: Penguin Books.
Sunder Rajan, K. (2006) *Biocapital: The Constitution of Postgenomic Life*. Durham, NC: Duke University Press.
Thacker, E. (2003) What Is Biomedia? *Configurations*, 11(1): 47–79.
Veraart, F., van Hooff, G., et al. (2011) From Arcadia to Utopia? In: Asveld, L., van Est, R., and Stemerding, D. (eds.). *Getting to the Core of the Bio-economy: A Perspective on the Sustainable Promise of Biomass*. The Hague, Netherlands: Rathenau Instituut.
Vermeulen, N., Tamminen, S., and Webster, A. (2012) *Bio-objects: Life in the 21st Century*. Farnham: Ashgate.
Wohlsen, M. (2011) *Biopunk: DIY Scientists Hack the Software of Life*. New York, NY: Current.
Žižek, S. (1999) *The Ticklish Subject*. London: Verso.
Zwart, H. (2009) The Adoration of a Map: Reflections on a Genome Metaphor. *Genomics, Society and Policy*, 5(3): 29–43.

# 2 Rethinking the biopolitical

> [B]eneath the dramatic and somber absolute power that was the power of sovereignty, and, which constituted in the power to take life, we now have the emergence [. . .] of this technology of power over the population as such, over men insofar as they are living beings. It is continuous, scientific, and it is the power to make live.
>
> –Michel Foucault, *Society Must Be Defended* (2003 [1976], 247)

## Introduction

'The power to make live', an awkward translation for the French concept of *le pouvoir de faire vivre*, is a fitting starting point for this chapter. It refers to the intervention in and regularisation of the life of the population but is also an apt description of how there are many 'species' of biopolitics that together can be seen as contemporary technologies of power with the ability to create new life forms that are alive and invented.

The passage quoted comes from the first pages of Foucault's lecture at the Collège de France, wherein the term 'biopolitics' was first used. He discusses the transformation of the power of the monarch over life and death, how monarchical sovereignty gave rise to the power to discipline the individual and the population over the course of the seventeenth and eighteenth centuries. As Foucault explains, the emergence of new technologies' power went from 'man-as-body to man-as-living being, and ultimately, if you like, to man-as-species' (Foucault 2003, 242). This 'if you like' is a direct reference to Marx's concept of species-being, derived from his 1844 manuscripts (titled 'De l'homme-corps à l'homme-espèce', sometimes rendered with 'être générique'). The word 'species' in the English version of this term suggests that it is a reference to biological nomenclature of types of life forms, which is not inaccurate, but in the original German it also denotes a general expression of a type or kind, *Gattungswesen* (see Fromm 1972).

Accordingly, Marx describes how labour is the means whereby man survives as an animal species and labour is the object of his life's activity. Not only is labour about satisfying the need to sustain existence, but it is also integral to the productive life of a conscious being (see Fromm 1972). Hannah Arendt explains the concept in her own language, referring to the 'devouring characteristics of biological life' and describing labour and its products as 'incorporated', consumed, and annihilated by the body's life process (Arendt [1958] 1997, 101–103). Ultimately, it is the absorption of individual lives into the life process of mankind that occurs through the 'liberation of the sheer natural abundance of the biological process' (ibid., 255). Foucault, in turn, introduced the terms 'biopolitics' and 'biopower' as part of an explanation addressing how 'man-as-species' becomes the object of increasingly complex systems of coordination of and intervention in general biological processes, 'covering the whole surface that lies between the organic and the biological, between body and population' (Foucault 2003, 253).

Foucault explains in detail the relation between the organic/body and biology/population as two sequential 'seizures of power' (ibid., 243). The first refers to the emergence of mechanisms, techniques, and technologies of power over the body 'in an individualizing mode' and as a 'whole field of visibility'. The seventeenth and eighteenth centuries witnessed gradual introduction of institutions such as schools, hospital wards, and workshops. Their introduction to the social fabric of public life imposed a system of subjection and gradual objectification of the individual bodies through surveillance, exercises, inspections, bookkeeping, reporting, and so on. The second seizure of power is not addressed to bodies but exercised over the life of the population. Rather than efforts to rule over a multiplicity of men, by controlling individual bodies, the late eighteenth and early nineteenth centuries saw the emergence of technologies of power directed at overall processes, such as birth rates, reproduction, fertility, longevity, sanity, and environmental conditions causing illness or malnourishment. Foucault called it 'a biopolitics of the human race' (ibid., 243–245).

Identifying these two seizures of power should not be interpreted to suggest that there is a breach between them, in the sense that one replaces the other or that either of them replaces the means by which power is exercised through a juridical system derived from monarchical sovereignty. Doing so is at the core of a common misunderstanding of the biopolitical, a viewpoint leading to claims that 'disciplinary institutions have swarmed and finally taken over everything' (ibid., 253). Rather, biopower came to exist alongside the power of the sovereign to decide who dies and who lives; 'it, in contrast, consists in making live and letting die' (ibid., 247). Sovereign right 'came to be complemented by a new right which does not erase the old right but which does penetrate it, permeate it' (ibid., 241–242). It implies a 'new right' that is not homogeneous; the emergence of a new technology of power does not replace its predecessors, 'but it does dovetail into it, integrate it, modify it to some extent, and above all, use it by sort of infiltrating it, embedding itself' (ibid.).

This insight is significant for our book and our claims made in the next chapters, particularly in that we are returning to the problem of sovereignty, *the power to take life*, as a theoretical horizon. Such an understanding of power may seem at odds with today's fully globalised world and with its criss-crossing circuits that allow easy travel of people, finance, media, lifestyle, and ideas across borders (flows facilitated by innovation in technologies, digital, and beyond). However, we maintain that examination of the problem of sovereignty is precisely one of the blind spots in today's discussion of new biotechnologies and bioeconomies operating across local, regional, and global levels.

Foucault insists in his lectures that the entire juridical edifice derived from the theory of right, the early-modern political doctrine of natural philosophy, is a problem, because it retains its hold over life even today when it is increasingly confronted by a heterogeneity of technologies of power. On one hand, it is a problem to be avoided, in the sense that sovereignty is a 'theory we have to get away from if we want to analyze power' and as 'the great trap we are in danger of falling into when we try to analyze power' (ibid., 34). On the other hand, the problem is that 'logically the emergence of technologies of power should have led to its 'complete disappearance' (ibid., 58). Yet it did not, and even in the completely globalised world of today, sovereignty persists as an organising principle, prompting the methodological question of how to examine the relationship between sovereignty and 'nonsovereign power' when the latter is 'impossible to describe or justify in terms of the theory of sovereignty' (ibid., 36).

Accordingly, this chapter establishes an understanding of the biopolitical that keeps the problem of sovereignty as its theoretical horizon and grounds the question of power. This is a methodological precaution leading to an analysis of the biopolitical that does not exaggerate what is new about the technologies of power in the context of life sciences surrounded by and constituted through futuristic affirmations and critiques. It is to this end that we start by describing the problem of sovereignty, or, rather, why sovereignty is a problem for our biopolitical present. This is also why we will first discuss Foucault as an 'anti-Hobbesian', which requires us to bring together his (later) texts on sovereignty with the (earlier) texts on the symmetry of life, language, and labour – the three dimensions to the underpinnings of the modern épisteme that, we will show, are being reconfigured in the emergence of new biosciences.

## The problem of sovereignty

Anyone who has a passing familiarity with Thomas Hobbes's theory of sovereignty will most likely have encountered the cover of *Leviathan*, showing a king whose body is made from countless figures against the backdrop of a city. The king wears a crown and holds a sword and sceptre, which illustrate the earthly and divine right to rule. The message is clear: only through monarchical power can the 'body politic' be unified, with the 'sovereign individuality' wherein royal power is the living embodiment ('I the state'). It is in this sense that also Foucault invokes Hobbes. His concern is not 'what the sovereign looks like from on high' but about the bodies that are constituted as subjects by 'power-effects'. The entire point, he explains, is 'to do precisely the opposite of what Hobbes was trying to do in Leviathan' (Foucault 2003, 28).

What does this mean, doing the opposite of *Leviathan*'s work? Aside from extensive analysis of that monograph in Foucault's lectures, there are various times that Hobbes is mentioned in his discussion of technologies of power. For instance, Foucault considers his studies of madness and punishment, both of which he uses as exemplars of the transgression of 'right' through power-effects, superimposed on the exercise of power to guarantee the cohesion of the social body (ibid., 33–37). Sometimes Hobbes is not mentioned explicitly but is the obvious reference point – for instance, in discussion of nuclear technology and the paradox of a sovereign who holds the right to kill and who, with excesses of sovereign power, could turn into 'the power to kill life itself' (ibid., 286–287). Hobbes features elsewhere too, notably in Foucault's lectures on the birth of biopolitics, wherein he discusses neoliberalism. Here his theory of the state is contrasted against the view that state power is maximised by creating freedom for the economy and against advocating for a state that continuously turns to the economy in the hope that 'its freedom can have a state-creating function' (Foucault 2008, 91–95, 312–313).

While there are many more examples, these do not suffice for describing Foucault as an anti-Hobbesian. After all, other philosophers are discussed frequently, and it is primarily in the lectures that there are constant and explicit references to Hobbes. That includes *The Order of Things*, which was written more than a decade earlier. In it, Hobbes is mentioned only once, in a discussion of economics, to point out Hobbes's view that it is the sovereign who authorises the circulation of money, giving it currency and animating its exchange (Foucault 2002, 195). Twice it is stated that law comes in many languages that 'by covenant or violence were imposed upon the collectivity' (ibid., 91). Entirely absent is the third domain, 'biology' (or, rather, 'natural history'), which would be suitable in light of the fact that he traces the origins of natural history to the era in which *Leviathan* began to circulate.

Yet this absence is not at all decisive. Most significant is how *The Order of Things* is set up; it is there that we see why later on Foucault turns to Hobbes so frequently. From the beginning, 'the problem of sovereignty' and, therefore, by extension, also biopolitics was at the core of his concerns. Taking the early and the later work on these topics together is what provides us with a point of view not previously explored.

## The classical space of representation

The introduction of *The Order of Things* discusses *Las Meninas*, the famous painting by Velázquez (ibid., 16–18), which was practically introduced to us in the same year as *Leviathan* (1656–57).

The painting is typically seen as depicting the household of the royal family of Spain. Foucault calls it the 'manifest essence' of classical representation because the work 'undertakes to represent itself in all its elements' (ibid., 16–18). The canvas (shown in Figure 2.1) presents a little girl in a white dress in the middle, surrounded by several other figures. She is not, however, the main subject of the painting – because of the reflection in a mirror of two faint images. This mirror reflects the Spanish monarchs and is surrounded by a self-portrait of the painter at work on a canvas, others in the royal household observing the monarchs, and an incidental observer looking through a door in the background. Crucially, this mirror is behind the figures depicted and does not reflect those who are directly

*Figure 2.1* *Las Meninas* or 'The Maids of Honour', painted between 1656 and 1657 by Diego Velázquez

in front of it, as seen in the painting. Instead, the mirror shows the Spanish monarchs as a reflection that cuts straight through the rest of the representation on the canvas. Furthermore, any observers of the painting are likely to be in the place where the monarchs seem to be standing. Foucault explains that the 'entire picture is looking out at a scene of which it is itself a scene' (ibid., 16–18).

The mirror is there to ensure that 'three points of view come together' – the gaze of observers of the painting, that of the painter as he was painting this scene, and that of the Spanish sovereigns who are reflected in the mirror (ibid.). Taken together, these points of view form

> a point exterior to the picture, an ideal point in relation to what is represented, but a perfectly real one too, since it is also the starting point that makes the representation possible. Within that reality itself, it cannot be invisible.
>
> (ibid., 16–18)

It is at this point that we can begin to rethink Foucault's perspective on Hobbes's power over life and how it is related to biopolitics. On one hand, biopolitics can be presented against the backdrop of the problem of sovereignty, which is to say that the edifice of right, once embodied by monarchs, remains at work in power's hold over life. On the other hand, it is in these passages wherein he explains the classical space of representation that the distinction from the classical conceptions of rights and sovereignty is most pronounced. The latter suggests that the point is not only that the mirror draws sovereign power into representation, showing a space of representation that would be impossible without the presence of the Spanish monarchs. That would simply imply an analogy between the court and a civilised society as imagined by Hobbes: a representation of the living embodiment of sovereignty. The crucial difference lies in the ideal point he mentions, a counterpoint that shows an 'essential void: the necessary disappearance of that which is its foundation' (ibid., 16–18).

Let us consider these terms in the context of the Hobbesian space, particularly his famous description of the human condition. Still today, Hobbes is ascribed the conventional political anthropology wherein man in nature is violent. In his terms, man's life in nature is characterised by 'continual fear, and danger of violent death; And the life of man, solitary, poore, nasty, brutish, and short' (Hobbes [1657] 1985, Chapter XIII). What this famous phrase illustrates is how Hobbes needed to invoke what surrounds the sovereign presence – for example, he presents life in nature as an escape from the misery of the human condition. It is on this condition, this foundation, that Hobbes can argue that man has decisively transferred his natural rights to the sovereign and concluded a social contract wherein sovereign power is and should be unlimited.

Many other natural philosophers, before and after Hobbes, sought to re-establish authority through such a state-of-nature theory. There are few who did not. Althusser, for example, mentions as the only exceptions the earlier example of Vico and later Montesquieu's histories (Althusser [1959] 1970, 25). Also Foucault says as much, placing Hobbes's theory of the state at the forefront in 'opening up of a domain of non-juridical social relations' and referring to Rousseau and Montesquieu as examples speaking to how history gradually became no longer juridical in its development, turning instead to the origin of civil society in the seventeenth and eighteenth centuries. What follows the establishment of sovereignty, as derived from a state-of-nature theory, is a history dealing with the relations between civil society as a given category and the authority of the state to the extent of entering into a 'completely different system of political thought' (Foucault 2004, 308).

This secondary history included a dismissal of state-of-nature theories, something that continues with more recent liberal theorists of the social contract, who denounce said theories as ahistorical methods and teleological foundations that are unnecessary (see Kymlicka 2002, 60; Rawls [1971] 1999). For us, however, these natural foundations are not merely a static and defunct juridical device but exemplary of the origin myth at the core of enlightenment philosophy that aids in understanding the classical space of representation. Not only does monarchical power sit at the centre of the judicial system of this era, but the exercise of power depends on ability to distinguish between order and disorder. The latter refers to man's rights in nature, anchoring, and setting in motion a juridical order that is a precondition for civility, as shown in the painting and as a Leviathan by Hobbes. The mirror is surrounded by the civility of the Spanish court as a reflection of the Spanish sovereigns, and, similarly, natural rights give rise to a social space that emerges out of 'one massive historical fact', which is the juridico-political theory of sovereignty (Foucault 2003, 34).

What, then, does it mean that Foucault sought to study power outside of the model of the Leviathan? The answer is visible in *The Order of Things*, where he approaches the problem of sovereignty contrapuntally, barely mentioning it while prioritising a painting that symbolises the transformation of how power is exercised and that relocates power to the space of representation that matches the sovereign's right over life and death. Yet what the painting pushes out of such a space of representation is the Hobbesian disorder. When it restores 'visibility to that which resides outside all view' (the sovereigns reflected in the mirror), this leaves undetermined the instability of its foundations, thereby establishing an essential void (Foucault 2002, 8).

Perhaps this is exactly as it should be within the context of 'the order of things'; we should keep in mind that the painting is shown as exemplary for the classical space of representation, and it is telling that we see the sovereigns who reside outside it. Yet we can now restore the state of nature to the space of representation, filling in for us the essential void that is needed to make representation possible. Today too, conventional theories of sovereignty are defined in terms of 'order', which is to say that 'disorder' continues to function as its foundation. Our aim, however, is not to describe disorder in terms of technology; attempting quite the opposite, we seek to take position against an overly historically naive drive to erase or normalise how the 'sovereign right' overflows and is saturated by contemporary technologies of power. The remainder of the chapter is dedicated to explaining how we will do this.

## The modern space of representation

What does such a position imply for our reflection on the current state of biotechnology, new bio-objects, (bio)economy, and biopolitics?

Technoscientific progress has allowed for the proliferation of new engineered biological beings, which can be described as experimental and open to new manners of appropriation, the capturing of economic value, and the like. However, whether globally, regionally, or locally, there are no uniform and clearly set rules, norms, or legislation by which newly engineered life forms and forms of life should, or even could, be brought under the power of the sovereign. Yet most of these novel objects are inevitably being made into subjects of the legislative powers of a sovereign state – that is, by current and emerging legislatures and political supporting institutions. Hence, the question is not simply one of order or disorder; it is one that requires understanding the 'void' described earlier, as it continues to be operational every time sovereignty is at stake, whether in the functioning of national governments (see Chapter 4), in relation to questions of global governance and global bio-economics (addressed in Chapters 5 and 7), or in rapidly changing biosciences (see Chapters 3 and 6).

Also today, it is the frame of the picture that holds conventional sovereignty together, with an essential void needed for enabling affirmation of specific ways of exercising power in their relation to the body politic. Examining the implications of extending the perspective to this subject matter entails a need to return to the theoretical and normative arguments in Foucault's lectures, which revolved around the need to 'bypass or get around' the problem of sovereignty (Foucault 2003, 27). This is directly applicable to how the following chapters address the constant work involved in recreating this 'outside' for the new species of biopolitics.

This chapter concludes with observations on how this outside can be perceived as a counterpoint, a vantage point from which to analyse contemporary technologies of power, but first some methodological considerations are in order. The aim is to take the step from examining the past (as Foucault did) to today's sovereign rights being permeated, saturated, and made possible by the embedding of new technologies of power. To that end, it is insufficient to understand merely that sovereignty is key to understanding today's predicament or to argue that species of biopower (and bioeconomies enabled by them) exist alongside conventional sovereignty. After all, it is the very novelty of life that is the speciality of new biotechnologies and bioeconomies, and we want to examine *how* it brings forward (into visibility) what is otherwise on the *outside* of representation.

Firstly, Hobbes's naturalism is an exception to the mechanisms via which the juridical formula that combined a state of nature with a social contract has lost much of its traction. The state-of-nature theories were immensely influential historically yet became optional or superfluous in liberal theories once the rule of law could be seen in terms of its own history (ever since Montesquieu). Hobbes's work is exceptional in that there has been no decline in the significance of his theory of sovereignty. Rather than forming the centrepiece of an argument about the monarchy as the living embodiment of 'civilized society', as opposed to a disorderly life in nature, Hobbes is inescapable today when one seeks to examine how power's hold over life constantly generates an 'outside' and an 'other' as a precondition for re-establishing sovereignty and the rights that are derived from it.

In this sense, the 'essential void' of the classical space of representation is with us now in many fields in which disorder is prioritised as a means to consolidate sovereign power, most obviously with issues of war, terrorism, crime, and so on. We could put it in the terms of *The Order of Things* thus: also today, the reflection in the mirror establishes how power is exercised, not only as a representation of sovereignty but by making visible the ideal point, the counterpoint, that was visible within the reality of the painting and that made the space possible. However, the conclusion as to the order of things could be interpreted as stating that Foucault argues that this position vanished at the end of the classical period. He writes that the 'symbolically sovereign' replaced royal power. Instead of actual monarchs, it is the observers of the painting who find themselves in the place of the sovereigns. However, they have become the 'enslaved sovereign' and an 'observed spectator' who occupies 'the place belonging to the king as was announced in advance by Las Meninas' (Foucault 2002, 340–351 see Chapter 2.3.1).

In precise terms, man:

> appears in the place belonging to the king, which was assigned to him in advance by Las Meninas [. . .] and as though by stealth, all the figures [. . .] (the model, the painter, the king, the spectator) suddenly stopped their imperceptible dance, immobilized into one substantial figure, and demanded that the entire space of the representation should at last be related to one corporeal gaze.
>
> (ibid., 340)

This extract is from the conclusion of the book, which reinterprets the introduction, wherein *Las Meninas* had been described as a representation of sovereignty in a classical space of representation. By the end of the book, following his detailed analysis of the history of science, he refers to 'one corporeal gaze' as having become characteristic of a modern space of representation. This gaze had become firmly established once natural history became biology, language became linguistics, and wealth became economics. What is corporeal about it is that these, as technologies of power, are established through practises and discourses, permeating the social fabric and the body politic.

### *The first methodological precaution*

A first methodological caution is that it would be a misinterpretation to reduce the scope of his book to an origin story: how all of the fields (biology, linguistics, economics) 'freed themselves from the pre-histories', leading to new concepts and methods that had been ignored for too long and that together shaped a purely scientific revolution (ibid., 274–275). Each field's formation is meticulously documented but as an element of parallel and mutually constitutive events that resulted in the modern space of representation and the corporeal gaze described earlier. For example, linguistics did not come into existence because of the way in which the grammatical system of Sanskrit was discovered, biology did not appear as a result of how organs were documented anatomically, and economics did not get it right once barter was subsumed by the economic role of capital. Rather, each of these new sciences had distinct methods and techniques that brought to visibility new objects that could sustain the flawless unity of knowledge. And its consequences were profound when considered in terms of what Foucault called the problem of sovereignty in later writings, as it implies that what had changed was knowledge itself.

Foucault is empirical in his research yet always ensures that his perspective does not depend on isolated events, individual facts, their interpretation, or any single field of science. For example, he discusses the re-categorisation of gills and lungs. It was well known in the Classical Age that both have as their function the respiration of species. Hence, there was nothing new about those functions being studied. However, the function of respiration did not govern, complement, or order the classical space of nature. Instead, Foucault emphasises, 'things will be represented only from the depths of this density', which in the instance of respiration refers to its function and how characteristics of very different species come to be seen as functional systems that exist one among many, each governed by others and seeking to govern life and nature (ibid., 274). He then refers to Cuvier's significance for the transition from natural history to the establishment of the scientific field of biology (in the last decades of the eighteenth century). He shows how the relationship between gills and lungs is transformed into one of many examples showing how Cuvier 'subjects the arrangement of the organ to the sovereignty of function' (ibid., 287).

Such examples demonstrate the transition to what is 'modern' about the space of representation of the nineteenth century. The development of our understanding of organisms had taken place within the confines of a 'discreetly preordained table of possible variations'. Breaking out of these given possibilities required that 'life' as a category become a 'space crossed by lines which sometimes diverge and sometimes intersect' (ibid., 295). The same applies to the other two fields. Wealth was no longer being conceived of as a system of equivalences, neatly listed in tables documenting its accumulation over time. Instead, 'production' and 'labour' establish constant measurement aligning the values of things, bringing 'wealth' into being, Foucault explains similarly that 'the sphere of grammar took place in accordance with the same model', replacing the form and sound of words in discourse with grammatical function and modifications that take place over the course of time (ibid., 305).

Taking these fields together is what shows the discovery of 'the general principle of an order', which is 'the constitution of a new space of identities and differences' (ibid., 296; see also 323).

Hence, there is no myth of origins or any other type of break with what came before. That applies directly to this book, as it shows how clearly we cannot shed the problem of sovereignty, make a clean break and move in a straight line to contemporary questions of life and technology, affirming a future that has freed itself from the past. If the corporeal gaze of the nineteenth century were 'it' – the end-point for the problem of sovereignty – we would have to concur with the misconception of Foucault's work as being part of a move further and further away from social wholes in a shift more and more toward the individual dissolved in an ineluctably advancing 'micro-politics of power' (see Said 1993, 336). We disagree with these diagnoses, based as they are on a misunderstanding or misreading of the concepts, isolating the lectures. His reading of the biopolitical should be seen in parallel to his earlier work, specifically how he describes the transition to the 'modern space of representation' of the nineteenth century as having come to pass. The problem of sovereignty is exactly that – a problem – not in the sense of a meticulously charted doctrine that regulates a social space but as procedures of law that are overwhelmed, crowded out, and prone to disorder in their relation to the biopolitics of the heterogeneity of the new mechanisms and technologies of power and life.

### *The second methodological precaution*

The second methodological precaution pertains to how our reading of the Hobbesian space is aligned with critical theory. For example, Max Horkheimer and Theodor Adorno appeal to, in *Dialectic of Enlightenment*, a comprehensive transformation of modern thought that is taking on the appearance of a 'nature that becomes visible in its alienation' as a second nature calling itself, as in pre-modern times (Horkheimer and Adorno 2002 [1944]; Horkheimer 2003).[1]

While this statement refers directly to the doctrine of natural philosophy as a whole, when considering its implications for technology we concur with Andrew Feenberg that it would be a misunderstanding to turn Frankfurt School critical theory into an attempt to revive a philosophy of nature in contrast to technology. Rather than suggest a road 'straight back to a teleological nature philosophy' (Feenberg 1999, 164), the point with this contrast is to provide an intellectual history for our self-understanding as subjects of technical action.

It is in this sense that Herbert Marcuse's notion of one-dimensional man is about technology as materialised ideology, imposing a system of domination that subordinates what human beings are and might be. What is one-dimensional is how 'the prevailing modes of control are now technological in a new sense [as] the very embodiment of reason for the benefit of all social groups and interests – to such an extent that all contradiction seems irrational and all counteraction impossible' (Marcuse 1991, 11). While this amounts to a claim that the social is being annexed by scientific rationality, it is important to be precise. As Feenberg explains, for Marcuse 'the blending of the technical and the social is not extrinsic and accidental, but is rather defining for the nature of technology' (ibid., 165). Feenberg states that 'just as technology is neither purely natural nor purely social, so the nature to which it is applied also confounds such abstract distinctions. Both are simultaneously causal mechanisms and meaningful social objects' (ibid., 165).

While such a position is compatible with our own, there is a noteworthy difference in regard of how power is exercised. In our view, the exercise of power, through sovereign rights and through technologies of power, is not solely about a right to invade, as it were, about how technology is 'colonizing the procures of the law' (Foucault 2003, 38–39). The Foucauldian perspective is characterised by transgressions that multiply in the meeting of the heterogeneous layers of constant confrontation. The second caution is, then, that the 'new space' we seek to describe is one of multiple transgressions, each of which can be examined against the backdrop of how Foucault's description of the sciences and the modern representation at every point makes explicit 'the necessary disappearance of that which is its foundation' (ibid., 38–39). Such necessity not only characterises the modern space of representation but generates a dynamic that can be investigated.

When we demonstrate that there is a contemporary space tied to the problem of sovereignty, we seek to show that it remains premised on 'a primitive and inaccessible nucleus, origin, causality, and history' that is buried 'deep down in the dense archaeological layers of the sciences' (ibid., 275). Hence, the 'second nature', as mentioned by Horkheimer and Adorno, incorporates what has been buried and returns to visibility within the context of life as a technological creation. On one hand, our analysis is conventional in how we will track this development over the decoding, recoding, and re-materialisation of life, approached as epistemic steps of (Latourian) translations. Each chapter touches upon scientific representation and the power it has to seem to be, for the rest of society, 'clear mirrors, fully magical mirrors, without once appealing to the transcendental or the magical' (Haraway 1997, 23–27; see also Shapin and Schaffer 1985, 39). On the other hand, this type of mirror provides a foothold for examining the problem of sovereignty and Foucault's framework for examining life, labour, and language in the nineteenth century.

Each subject treated in this book shows a different dynamic but thereby can be seen as a component, a step, and a precondition for pursuit of the overall aim, which is to examine what we referred to in Chapter 1 as 'metacodes'. As an element of methodology, this concept refers to the Foucauldian project in its aspiration to approach the biopolitical without reducing it to an otherwise bewildering heterogeneity of discourses, techniques, circuits, and types of research and governance.

## Conclusion

The conclusion points toward some of the theoretical implications of rethinking the nature of sovereignty and its relations to biopolitics. A presentation of these before the analysis in the remainder of the book must be abstract by its very nature but could have been made much more concrete via imagination of the various chapters in line with Picasso's fifty-four reinterpretations of *Las Meninas*.

Regrettably, this proved to be impossible or, rather, possible only on condition that we pay for the use of a copyright that either is dubious (because Picasso made reproduction of a work that is centuries old) or undermines the doctrine of 'fair use' (or 'fair dealings' in Britain). Other authors have described this particular copyright claim as an example of how intellectual property is a socially constructed concept (George 2012), and previous nationally based publications in books by publishing companies have acknowledged that making critical use of the painting implies that doing so falls under fair use (e.g., Deibel 2008). Here, the argument for payment revolves around printing for global distribution. While the only requirement in most jurisdictions is that the use should not undermine the ability to profit from a copyright, one can appeal to the strictest conceivable copyright claim in cases of global publication, apparently without even the necessity of a particular legal system being proved relevant. Rather than acknowledge such a copyright, we chose to leave out the picture, which can be readily found on the Internet.[2]

The image that should have been visible is one of Picasso's variations.[3] Like the others, it is a composition filled with what are almost Cubist figures, which are nonetheless easily recognisable as the princess, the maids of honour, the incidental bystander, and the reflection of the sovereigns from Velázquez's original. In this version, a mirror, a dog, the door, and the painter (who is shown as a huge figure with a face that mirrors only himself) constitute the space of representation. Each of them is in the same place in the composition as in the original but with a crucial difference: the mirror is not superimposed. It resembles the original even though the reflection of sovereignty has become only a faint hint of power, and yet it is still able – by way of its Cubism – to draw the observer into the representation. These interpretations by Picasso, unlike the 'enslaved sovereign' that Foucault concludes with, foreground that it is possible to incorporate the distance in between the work itself and the original. It is the representation of this 'in between' that is crucial; it can be imagined as analogous to the formation of the modern space that precedes it, transgressive in its relation to the myth of origin, multiplying a disappearing foundation integral to power's hold over life.

The comparison gives an impression of how we are to approach the subjects we will discuss in the subsequent chapter. The painting shows, through a Cubist rendering of the original, how the problem of sovereignty can be re-articulated, or how it can sustain a relation to the classical space of representation. We began by pointing out that the contrast between biopolitics and conventional understanding of sovereignty is misunderstood when seen as a breach or in terms of epochal interpretation. A better interpretation, we hold, is to see the transition from the classical to a modern space of representation as visible and continuous transcending of the limits of rights and power that become invested in real and effective practises, circulating, and functioning by passing through individuals, constituting subjects and subjectivity (Foucault 2003, 28). It is this transcending or even transgressing that characterises the implementation of 'right' through the 'multiple relations of power [that] traverse, characterize, and constitute the social body' (ibid., 48).

It is exactly to capture this transgression that we diverge from the Marx-Foucault synthesis that is characteristic of much of the critical literature on the life sciences and related bioeconomies. What remains the same is that we also study how technologies of power that regulate the population are coupled with the 'dynamics of labour and commodification that characterize the making and marketing of such entities as industrial and pharmaceutical bioproducts' (Helmreich 2008, 464; see also Chapter 8). This combination of technology and political economy fits perfectly within our framework, and as a critique it is based on Marx's species-being as the theoretical representative of the moment when critique began actively seeking to 'frustrate any attempt to isolate a state of nature separate from society' (Thacker 2005, 38; see Deibel 2009). Simultaneously, this is one of many of the discarded theories, belonging to the classical space of representation, that form the disappearing foundation that we need to understand methodologically. Each of these theories gives us its own distinct perspective on the essential void that characterises the interplay of technologies of power. Hence, one cannot separate political economy from the study of science and technology or prioritise one over the other. Rather, we seek in our approach to sustain symmetry – among life, labour, and language as well as among different types of technologies of power, or between technology as a variable as opposed to economics, or the language of life as opposed to its biology.

To do so, however, requires a biopolitical horizon that is anchored in the problem of sovereignty. This is not self-evident, as it implies an archaic doctrine in discussion of technologies of power, which often adopt futuristic language. The move we seek to make in each chapter is captured already in the two paintings. On one hand, we have Hobbes's foundation, with *Las Meninas* questioning what is visible in the mirror via the non-visible disorder that surrounds it. What is left outside, consigned to the disordered space, is the non-conforming elements of modernity, anxiety, and monsters and cultural catastrophes that threaten social order, thereby giving sovereign authority its legitimate grounding. On the other hand, we see a Cubist rendering of the modern space of representation, a new space of multiplicity that remains closely tied to the classical space of representation. At the same time, its recognisable contours show us the new space of representation that is already here and coming after it.

Contemporary equivalents of states of exception and non-conforming elements causing modernist anxiety are easy to find, as a global state of war populated by failed states and terrorism but equally in declaring war on drugs, disease, or hunger. Even appeals for food security and safety are likely to speak to various insecurities, appealing to the sovereign to act in response to the risks of genetic engineering or climate change or to sanction, secure, safeguard, and monetise the genetic wealth and resources in the sovereign's territory (see Chapters 4, 5, and 7; see also Deibel 2013). Theoretically we draw on a similarly diverse repertoire. For example, Hobbes can easily be replaced as the basis of sovereign legitimacy by drawing on how Rousseau turned the state of nature into a constitutive fiction (see Chapter 5) or anchoring to Grotius's ideas of international law as applied to the natural world on the outside of civilised societies (see Chapter 6). These and other political theories are readily compatible with the thoughts of Hobbes and of Foucault as an anti-Hobbesian, disturbing very little within the classical epistemological arrangement while allowing us to rethink the disappearing foundation of the classical space of representation as the 'essential void' that continues to characterise how power is exercised.

To conclude the chapter with a demonstration, we can take one of the most obvious candidates, Locke's theory of labour. What is obvious is that Locke's theory is directly relevant for questions of biotechnology and particularly in regard of how difficult it has become to identify property and ownership with man's possession of his body and its strengths. It is with Locke's theory that the body and its labour became 'the quintessence of all property' (Arendt 1997, 112–115), although this did not last for very long. Obviously, today there are constant disputes over whose labour led to a particular breakthrough invention and hence who should be awarded the patent and associated privileges (see for example the CRISPR-cas9 genome editing technology patent debate between the Broad Institute in Cambridge, Massachusetts, and the University of California over intellectual-property rights to the potent technology). Similarly, identification of an invention with an individual or even a group of individuals is hardly ever straightforward in the life sciences. In this sense, bio-economics can easily be seen as a case highlighting how arbitrary it has become to hold that labour is derived from how individuals are in possession of their body. No such body will be found when ownership is contemplated in relation to seeds, organs, genetic wealth, and the process of biological reproduction as informatic resources to be acted and interacted with (see Hardt and Negri 2004, 187). In sum, the situation we are faced with is a constellation of complicated and precarious claims, rights, and demands that need to be established in relation to processes of gene expression and regulation and also sequences that can be cut, spliced, and transcribed.

Returning to the problem of sovereignty, however, we are not done upon having denounced Locke's labour law and the paradoxes of contemporary ownership. Man's possession of the labour of his body is not only a result of Locke's famous dictum about the 'innate rights of life, health, liberty and possessions' (Locke [1689] 1970, §6). Rather, the disappearing foundation lies in this: in Locke's state of nature, man is as insecure as with Hobbes's formulation, but for a different reason. While he is living freely and is healthy, the fruits of his labour are unsafe. It is to protect his possessions that he leaves nature, and, hence, the social contract does not grant the sovereign the authority to take his property away. As he never gave up his natural right to property, the sovereign has the duty to safeguard ownership. What difference does such an understanding make, aside from introducing different types of transitions between nature and society? Each instance gives us a distinct type of transgression that is 'symbolically sovereign' and that can be restored to visibility. In our case, we can examine the many relationships between information and the biopolitical, as transgressions that converge in how life and nature undergo an 'involution' much as space did – from the expansive to the interior aspects of the natural world, from an extensive to an intensive project of exploration, which is 'nowhere more evident than in the emergence of contemporary biotechnology' (Parry 2004, 49).

In other words, the technologies of power that are constitutive of man-as-body and as a living species are turning into sources of its body anxiety experienced in recognition that a dream of a self-sufficient autonomous, nearly immortal body loses much of its meaning when realised as an informatic body that is contingent and unstable, in need of constant monitoring, programming, and engineering (Thacker 2003, 89). While we decry the notion of there being a positive condition of a transhistorical belief in a human and modern subject that might be displaced at any time along with its environment (as a human and modern subject with a defined beginning and ending), we seek to enrich the analysis of biopolitics as comprising multiple political theories of sovereignty in the classical space of representation.

By doing so, continuing with the theme of the body and diverging elements of power in subsequent chapters, we seek to think across the many distinctive transgressions of informatic ways of thinking about life and nature, each of which is visibly saturated with paradoxical and critical terms, conditions, and myths of origin surrounding the problem of sovereignty. In summary, each of the chapters engages directly with the empirical heterogeneity, yet, we claim, in doing so brings us to a shared condition – to a larger picture of the biopolitical that we can understand in terms of 'metacodes of life'.

## Notes

1 The wording is translated directly from the German version on which the citation is based (see the title page of the cited volume). To be precise, '*natur, die in ihrer Entfremdung verhehmbar werd*' literally refers to 'nature, that in its alienation becomes visible', not 'nature made audible as estrangement'.

2 The observation that this book qualifies as 'global commercial distribution' was the only argument offered in response to our observation that 'fair use/fair dealings' allows for a reproduction in support of a critical discussion of the painting. No further arguments were provided by the agencies involved, nor was any further attempt made to explain why fair use is not applicable nor why a copyright on this particular work exists at all. Usually, critical use in books is allowed under the fair-use doctrine if that use does not undermine the profitability of the copyrighted work. Here the only argument seems to be that there might be jurisdictions somewhere in the world in which there is no such interpretation of fair use. It is our (the authors') problem to figure this out, not theirs: we were simply forwarded to a royalty collection agency (interpretation based on personal e-mail communication by Deibel with the copyright-holders and related collection agencies). These types of practises are clear instances of contemporary equivalent to the enclosure of the commons (see Chapter 4) and should not be accommodated.

3 For any of the works that we might have used, search for Meninas and Picasso. The one we would have used is the most obvious one, found here: https://en.0wikipedia.org/index.php?q=aHR0cHM6Ly9lbi53aWtpcGVkaWEub3Jn3dpa2kvTGFzX01lbmluYXNfKFBpY2Fzc28p (accessed on June 2017).

## References


Althusser, L. (1970) *Politics and History: Montesquieu, Rousseau, Marx*. London: Verso.

Arendt, H. (1997) *The Human Condition*. Chicago: University of Chicago Press.

Deibel, E. (2008) Recoding Life in Common: A Critical Approach of Post-Nature. In: Ruivenkamp, G., Hisano, S., and Jongerden, J. (eds.). *Reconstructing Biotechnologies: Critical Social Analyses*. Wageningen, The Netherlands: Wageningen Academic Publishers.

Deibel, E. (2009) *Common Genomes: On Open Source in Biology and Critical Theory Beyond the Patent*. PhD dissertation. Available at: http://dare.ubvu.vu.nl/handle/1871/15441 (accessed on December 2016).

Feenberg, A. (1999) *Questioning Technology*. London: Routledge.

Foucault, M. (2002) *The Order of Things: An Archaeology of the Human Sciences*. London: Routledge.

Foucault, M. (2003) *Society Must Be Defended*. New York, NY: Picador.

Foucault, M. (2008) *The Birth of Biopolitics*. New York, NY: Palgrave Macmillan.

Fromm, E. (1972) *Marx's Concept of Man*. New York, NY: Ungar.

George, A. (2012) *Constructing Intellectual Property*. Cambridge: Cambridge University Press.

Haraway, D.J. (1997) *Modest_Witness@Second_Millennium.FemaleMan©_Meets_OncomouseTM: Feminism and Technoscience*. New York, NY: Routledge.

Hardt, M. and Negri, A. (2004) *Multitude: War and Democracy in the Age of Empire*. New York, NY: Penguin Press.

Hobbes, T. (1985) *Leviathan – Edited With an Introduction by C.B. Macpherson*. London: Penguin Books.
Horkheimer, M. (2003) *Eclipse of Reason*. New York, NY: Continuum.
Horkheimer, M. and Adorno, T.W. (2002) *Dialectic of Enlightenment*. Stanford, CA: Stanford University Press.
Kymlicka, W. (2002) *Contemporary Political Philosophy: An Introduction*. Oxford: Oxford University Press.
Locke, J. (1970) *Two Treatises of Government: A Critical Edition With an Introduction and Apparatus Criticus by Peter Laslett*. Cambridge: Cambridge University Press.
Marcuse, H. (1991) *One-dimensional Man: Studies in the Ideology of Advanced Industrial Society*. London: Routledge.
Parry, B. (2004) *Trading the Genome: Investigating the Commodification of Bio-information*. New York, NY: Colombia University Press.
Rawls, J. (1999) *A Theory of Justice*. Oxford: Oxford University Press.
Said, E. (1993) *Culture and Imperialism*. New York, NY: Vintage Books.
Shapin, S. and Schaffer, S. (1985) *Leviathan and the Air-Pump: Hobbes, Boyle, and the Experimental Life*. Princeton, NJ: Princeton University Press.
Thacker, E. (2003) What Is Biomedia? *Configurations*, 11(1): 47–79.
Thacker, E. (2005) *The Global Genome: Biotechnology, Politics and Culture*. Cambridge, MA: MIT Press.

# 3 Read, write, standardise[1]

## Skin

> In the electric age we all wear all mankind as our skin.
>
> – Marshall McLuhan, *Understanding Media* (p. 47)

In the late 1960s, Marshall McLuhan proclaimed that in a near future – in a future that resembles to a terrible degree our global contemporary present – the narcissist tendency of humankind to continuously seek the most perfect image of itself would drive the species toward existential discontent. According to McLuhan, this narcissist appetite would drive a desire to seek a constant 'upgrade of the self', which, in turn, would lead to an anthropological narcosis manifested through a numbness of the primary senses. This numbness, then, would be a consequence of the different technologies developed to extend our senses beyond what the body normally would be capable of doing, experiencing, and expressing. His underlying suggestion was that technologies create and operate within a space of the new media that is a reaction to a deep existential discomfort, experienced through the pains and sicknesses of the body – an anthropological experience of deep-running dissatisfaction. For him, diverse technologies developed to enhance our senses and the functions of the body would equate to technological prostheses and amputation of the primary bodily functions – and, more generally, our intentions as embodied beings.

The flickering face on the water, as the story goes, was so captivating that all else in life seemed meaningless for Narcissus. A similar tendency of autoeroticism can be witnessed in the bioscience research pursuing the ultimate truth about various species via their genetic makeup, in the technoscientific enterprises driving the development of the bodily and informatic capacities of our species body. And the face is not just reflected but also subjected, such that for the first time we become aware how we, as individuals and species, are equipped with a large number of technological **extensions** of the body. At the same time, and beyond extensions, we also have developed externalised **intentions** placing our identities outside our embodied being – our mind is being increasingly redistributed in its material and informatics senses (Hutchins 1995), while our bodies are extended with numerous technologies of life.

This dual recognition stems from the fact that biological bodies themselves are both a medium of coded functions that run inside cells, guided by instruction sets (in DNA), AND a source of numerous unknown bits of information to be decoded for new biotechnological pursuits

> [b]y putting our physical bodies inside our extended nervous systems, by means of electric media, we set up a dynamic by which all previous technologies that are mere extensions of hands and feet and teeth and bodily heat-controls – all such extensions of our bodies, including cities – will be translated into information systems. Electromagnetic technology requires utter human docility and quiescence of meditation such as befits an organism that now wears its brain outside its skull and its nerves outside its hide.
>
> (McLuhan 1995, 57)[2]

The anthropocentric figure of Narcissus, and *Homo sapiens* as its own species body, thus became quiet reflections of an extended organism, an extended sense of being in which **we all wear humankind as our skin**. Skin – as the outer physical boundary of our bodies and embodied individuality – still marks the difference between individuals. New technologies are generative in relation to the media they are embedded in, but this generative vitality is an outcome of our embodied human experience that has a profound capacity to reach out, extend, and redistribute our sensing capacities through information systems, even though the body itself might be physically delimited by the skin (Bateson 1972). At the same time, **skin** serves as the boundary site for inscribing in our bodies new attributes, capacities, and dreams of the whole of mankind, mediated through global information networks. Skin works as the physical human–world interface between the individual and mankind, body and information, function and intention, now newly articulated in relation to an 'extended nervous system' at global scale, as McLuhan saw it – the Internet.

McLuhan's proposition as to how media and their constant transformation beyond known formats and circuits – text to film to electric lighting – concentrated on rethinking what we should consider to be 'information' and how bodies will be capable of carrying, communicating, or transmitting any of that. With the biomolecular revolution (Kay 2000), the cybernetic movement of 'in-formation' during the important decades of the mid-twentieth century (Hayles 1999), and the idea that DNA is information contained in the medium of bodies – code of life to be not only decoded but also encoded and now recoded (Thacker 2010) – our bodies have become transformed into several distinct media formats that together Eugene Thacker calls 'biomedia'. For him,

> biomedia is a constant, consistent, and methodical inquiry into the technical-philosophical question of 'what a body can do' [. . .] particular mediations of the body, optimizations of the biological in which 'technology' appears to disappear altogether [. . .]. The apparent paradox of biomedia is that *it*

> *proceeds via a dual investment in biological materiality, as well as the informatic capacity to enhance biological materiality.*
>
> (ibid., 52–53; italics in original)

The enhancement of biological materiality is, on one hand, driven by primary and practical goals of **healing** that are rooted deeply in the core of the *anthropos* itself. This also renders new information about the body valuable, both as direct interventions with the body and indirectly in the practises by which the body can be healed, re-balanced, or enhanced. At the same time, the question of the 'body' and what it can do becomes newly articulated and pertinent to a vast array of industries, institutions related to the generation, processing, and application of the new biomedia. Here, the question is at once pragmatic and philosophical, and value becomes recast in terms of economic and experiment-oriented idioms.

While the informatic paradigm of life suggests that 'life itself' and natural beings are all born out of evolution, it suggests also that evolution and diversity of life at their core are fundamentally an error,[3] a copying error that generates individuals through variation brought into the code of life. But when 'life' is recast as information evolving through error and mutations in the code, also ideas related to biomedical interventions with the body – for example, healing and enhancement – are re-articulated in terms of how to go about these interventions. Georges Canguilhem, a French historian of science and philosopher, pondered this question in the 1960s era and asked:

> How is evolution to be explained in terms of genetics? The answer, of course, involves the mechanism of mutations. One objection that has often been raised against this theory is that many mutations are subpathological, and a fair number lethal. So the mutant is less viable than the original organism. To be sure, many mutations are 'monstrous' – but from the standpoint of life as a whole, what does 'monstrous' mean? Many of today's life forms are nothing other than 'normalized monsters', to borrow an expression from the French biologist Louis Roule.
>
> (Canguilhem 1994, 318)

Understanding living beings as producers, transmitters, and receptors of vital information borne by a copying error – and resulting in greater diversity of functions and applications, some lethal and some just monstrous – allows one to focus on the question driving current biomedical practises that go beyond the idea of expansion of the senses and proliferation of biomedia. These drivers manifest themselves today in the global efforts centred on reading, writing, and optimising 'the code' (for example, embodied today as versions of optimised functions manifested through novel synthetic life forms; see Chapter 6). The idea of reading, writing, and optimising is related to that of the 'normal' and 'baseline' performance in relation to the errors in the code at the individual and species-body level – and to the question of the monstrous as Canguilhem phrases it. But this new business of optimisation requires a way to speak about living beings through new forms of biomedia, to operationalise and transmit information in the right bundles to enable biotechnologically assisted intervention, along with a way to pool this information together so that we can 'wear all mankind as our skin'. Let us explain.

## Tongue

Biology, as an early-nineteenth-century epistemological configuration, has been obsessed with objects of various sorts throughout its history. Examples are fleshly objects, physiological structures, and evolutionary forces. From the middle of the twentieth century, with the discovery of DNA and with the development of cybernetic approaches, biosciences have increasingly turned to informatics and patterns as the sources of discovery. And in a shift that began in the early 2000s, the life sciences have become dependent on computation power to crunch large datasets and thereby identify and characterise phenomena related to underlying biological mechanisms, pathways, and systems. As biologists 'think bigger' than ever before, their work's appetite for data grows ever more voracious (Hadley 2004). It demands gathering and administration of large collections of samples and related data; however, this is not an endeavour for individual biologists, single projects, or smaller research institutions, nor can it be, on account of the high costs (in time, technological resources, and funding) involved. Biology in itself has become 'big science' (Vermeulen et al. 2013), with institutional collaboration and geographically distributed forms of endeavour.

Accordingly, biomedical researchers have themselves started to envision how science of biology is dependent on a 'science of biobanking', a science that is 'meta' to the actual bioscience itself, one that acts as an enabler for the bioscientific discovery through biobanks and contained therein. And this is through not only *a* biobank but *all* biobanks and their data on the human condition on global scale. Biobanks hold biomedia – organs, tissues, samples of all kinds, data, and diagnostics (increasingly in digital format) – and by doing so, they take part in shaping the global vision of informatic *anthropos* and its condition of being, the metascience of biomedia.

For example, in an article titled 'Toward a Roadmap in Global Biobanking for Health', published in 2012, 25 well-known biomedical researchers from across the EU described why a metascience is needed (Harris et al. 2012, 1110):

> The science of biobanking itself is as important to develop and fund as the science that uses biobanks. Because the science of biobanking is very closely linked to the development of an enabling infrastructure, it requires scientists to work more closely with each other and with funders than has historically been the norm in biomedical science. In this, biobanking resembles more the situation typical for large-scale physics, which is characterized by close collaborative, pre-competitive relationships among workers in the field to construct, develop, and maintain the necessary research infrastructures while embracing healthy competition in the undertaking of both hypothesis-based

> and free research using these infrastructures. Developing this kind of culture and spirit will be essential for the success of large, cross-institutional biobanking efforts.

The key idea is to generate a large-scale infrastructure of biomedical samples that are made readily accessible so that biomedical research can develop further. Thus, the vision of large-scale bioresearch infrastructures, of connected biobanks, also introduces a vision of a new social and institutional organisation, or, as the quote puts it, a new culture and spirit of research. Biobanks have become the source and site of many a biomedical datum, demanding that individual scientists, funders, and institutions join efforts and work closely together, though in a state of 'healthy competition' – a research paradigm that is contextualised in liberal, individualistic, and capitalist culture. This new culture and spirit are found at the heart of the new science enabling bioscience: the science of biobanks, or the 'metascience' of biomedical discovery.

This science of bioscience, complete with the enabling biobank infrastructure operating under a new culture and spirit, is itself dependent on one very important feature: interoperability. To stress this point, the twenty-five researchers developed a ten-item list of considerations underpinning a roadmap for global biobanking. The first and most important of these was a call to 'foster biobank interoperability through the development and maintenance of enabling technologies, procedures, networks, and compatible ethico-legal frameworks' (Harris et al. 2012, 1110). Interoperability is key, because within the 'big science' paradigm, collaboration requires mutually agreed policies, standard operating procedures for sample collection, and common standards for information's representation and sharing. In other words, collaboration requires a common language and set of core practises, with shared epistemic and ontological commitments underpinning the common research infrastructures now being developed around the globe. *In other words, a mother tongue for biological samples across studies, collections, and collaboration boundaries is needed.* The urgent need for a native common language seems at the outset to be a minor feature in the midst of larger efforts involving 'real science', informatic-system development as a reflection of organisational policy renewal. However, at a closer look, it becomes clear that this is the definition of a common language, a mother tongue for biobanks that enables all of the science of biobanks to evolve and makes large-scale biobank-based bioscience possible in the first place.

Life scientists and policy makers have been aware of this challenge for a long time. The European Science Foundation reflected on this problem in 2008 and wrote, in a report on population-wide biobanks, that

> [g]ood, inter-operable Information Technology (IT) systems are required so that information contained in the different datasets can be adequately mined by integration or, at least, interfacing, and efficiently linked to relevant information from other sources. The fact that many biobanks or biobanking networks use different IT platforms and different message formats

> and terminologies represents a significant obstacle to communication with, within, and between biobanks.
>
> (European Science Foundation 2008, 8)

A native language allowing biobanks to talk with each other at the technical level (for messaging formats, terminology, and databases) has been missing for a long time. Instead, we have many competing ways to represent biological samples and related data. Over the years, there have been attempts at various levels to bring together one standardised way to talk about vital matters. These efforts have ranged from regional and international organisations'[4] promotion of principles such as quality standards to science and infrastructure initiatives[5] offering large-scale networks of partnership for bioresource tools and databases:[6] tools and repositories for data. However, the efforts to standardise (or even to 'harmonise' the ways of talking about biological samples, genetic databases, or any other form of data belonging to biobanks) have encountered severe challenges related to making things more interoperable.

When interviewed on the topic, a leading scientist working on some of the largest coherent biobank datasets at a national health institute in the Nordic region described the challenge thus:

> There are so many institutions, databases, practices around, and almost as many persons wishing to harmonise the biobanks and other repositories into a single effective way of working together that I don't know what's going to come out of it. I've been saying that somebody ought to **harmonise** the harmonisers!
>
> (Interview in December 2014)

To illustrate how language – and phenomena related to differences in languages between cultures, computers, and domains of the 'bio' – is inherently tied to the way in which large-scale biobank research infrastructures work, we will review the development of the standard BBMRI information model called Minimum Information about Biobank Data Sharing, or MIABIS (Norlin et al. 2012). The model is intended to provide a standard for integrating biobanks across Europe into a common trans-Europe virtual network of biobanks. It is a working implementation of the proposed biobanks standard, the first attempt to provide an informational backbone for the large-scale biomedical infrastructure platform envisioned by European research policy.

MIABIS was developed to facilitate the Biobanking and Biomolecular Resources Research Infrastructure (BBMRI) envisioned as becoming the key scientific research infrastructure for European life sciences. The pan-European BBMRI vision emerged from the political recognition that keeping up with developments elsewhere, most notably in the United States, necessitates integrated European research. Development of a pan-European infrastructure is driven by the vision of bringing together geographically dispersed research communities and distinct life sciences disciplines (as in biology plus medicine) with the aid of a specific branch of information science fundamentally informing radical transformation of the research practises: bio-informatics (European Strategy Forum on Research Infrastructures 2006, 23).

Large-scale infrastructures are notoriously difficult to build and to manage, and governing them is rife with challenges. Large-scale infrastructure aimed at translating political decisions and statutes into reality is even trickier to implement. Building large-scale infrastructure such as the BBMRI poses a number of challenges for bio-informaticians implementing work toward the political goal, most importantly in the building of a technical platform that could successfully integrate the disparate information models already in use. The integration must link sample collections and studies, deal with natural languages' barriers and differences in lexicons, and address legal provisions related to protection of privacy.

What the analysis that follows shows is that it is not science but 'metalanguages' that render it possible to standardise the language of life (samples, medical records, and other information) throughout the EU. Life is primarily known not through DNA, biological samples, or any other material corporeality but through a metalanguage that ties together the complexity of life forms and forms of life as known by biomedical sciences. The native tongue of the biosciences – the 'metalanguage' drawing large-scale biobanks and other biomedical platforms together – has a very particular bearing with regard to the original living bodies. The native tongue of any informatics platform directly informs the way biomedical bodies are shaped, made to perform, and known.

## Body

> This integration standard will make a great contribution to the discovery and exploitation of biobank resources and lead to a wider and more efficient use of valuable bioresources, thereby speeding up the research on human diseases.
>
> –Roxana Merino-Martinez et al. (2016)

Developing a native tongue for biobank harmonisation and standardisation depends on how this language correlates with the bodies it depends on. The question about the body – the species body, the sovereign or national body, and the body of the individual and of the institutions it inhabits – has been one of the vexing questions drawing together biology, the social sciences, economics, and various threads of psychology and philosophy. Population and its characteristics, the governance of multiplicities of bodies, the perception of the body, and its institutional inscriptions (through productive qualities, gender, race, class, sexuality, or biomedical properties) have all been key concerns for anthropological enquiry since the beginning, although systematic treatments have been 'sporadic throughout the history of the discipline', as Margaret Lock (1993, 133) notes.

Bodies are the central object and problem for biomedical science and biobanking efforts, mainly for the same reason that they have been for researchers in the other fields. As soon as one starts to describe the qualities of a body, it dissolves and translates into several other concepts: skin, organ, cell, molecule, DNA, and single-nucleotide polymorphism (SNP, a type of genetic marker). There are many alternative epistemologies of the body within biomedical sciences, rendering bodies ontologically unstable objects and subject to objectification within particular and localised practises such as diagnostics, correction, and care (Mol 2002). Simply put, there are no common definitions circumscribing what a body consists of and what a body should be or do on a bio-informatics-driven biomedical research platform. Within biobanking, for example, one can easily distinguish clinically based and population biobanks. The former consist of a large mass of population data and are geared for epidemiological research (derived as they are from bodies that can be used as markers of predisposition and population-level exposure to external risk factors), while the latter focus on biomarkers for disease (e.g., derived from bodies used for prognostics, predictive analysis, and sources of personalised medicine) (Harris et al. 2012).

To develop a universal language that speaks in the name both of populations and of individual bodies, external risk factors, and pathogens, one must, however, consider also other kinds of bodies. There are sovereign and national bodies at stake here, and the language of biobanks needs to capture these too. For a common European research infrastructure such as the BBMRI, one of the key challenges in dealing with the multitude of bodies – national populations of all kinds – is their political representation through language.

First, the bodies biobanks hold are represented by natural languages that might not be translated easily one to another. Swedish and English, for example, have different words used to describe what the objects in a given institution are. Second, these bodies – the 'specimens' or 'samples' – and, at a more precise level, the organs, the fluids, and the information attached to them might reside not in a legal entity called a biobank but in a repository for biomedical samples. The leader of the standardisation efforts in Europe, who leads the Swedish bio-informatics group working at the Karolinska Institute, commented on this issue in the following manner:

> One of the issues in Europe is language. You see, the terms and definitions you use for biobanks are a sort of a problem because there are so many different borders with BBMRI and everyone [has] their own terminology that differs from each other. So we have created a lexicon. The current version default [for the database information standard] is English, but what we did was to provide the information and the service in 10 different languages.
>
> (Interview from 21 November 2013)

If the concept of biobank in each EU language denotes a specific local, country-level configuration of sample collections, data, and study information, how can a scheme make sense of them all? The Swedish bio-informatics group explicitly anticipated the issue of terminology and created a lexicon of biobank-related concepts, providing its own translation of these between ten European languages. Thus, standardisation between languages was deemed the first translation required.

However, this standardisation can only occur within the BBMRI context. The definition of biobanks is not only a matter of providing inter-language equivalence for the EU's various medical communities. This is because the concept of biobank can differ in content also within the various epistemic communities in a particular member country. One medical community may have its own working definition of 'samples', while a similar community in a neighbouring country might entertain another.

These new lexicons, crucial as they are to the biobanking practises, transform the way in which bodies and the discursive institutions they inhabit are articulated at local levels. Establishing a language that is universal for inter-cultural discourse about the objects held in biobanks – for example, samples, fluids, and information – cuts across borders between national lexicons and between states.

The state and its sovereign power to name, intervene in, and organise its population and individual bodies is disrupted at the level of universal language. A powerful two-pronged move in favour of 'interoperability', powered by an agency proceeding from the scientific vision of a common, standardised European biobank database, was prepared by a group of bio-informaticians in Sweden. The database and the lexicon, (see Figure 3.1) as its ontological grounding, transform the European

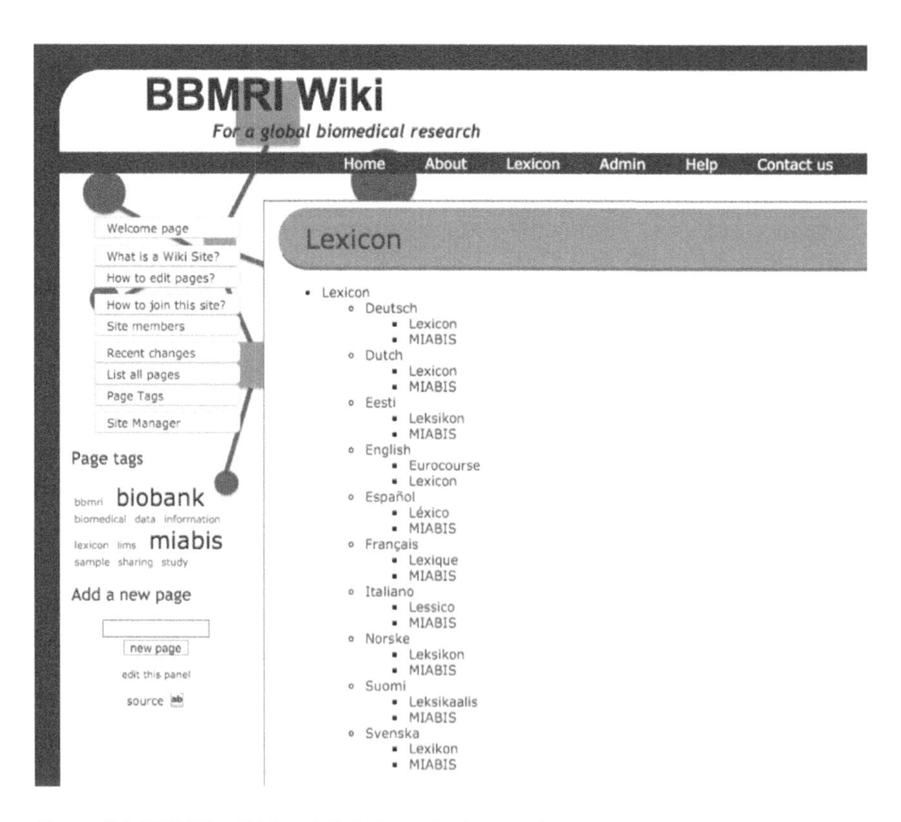


*Figure 3.1* BBMRI wiki for global biomedical research

bodies into standardised samples. They become organs removed from their political bodies and rerooted in the technical discourse of the European biobank. This is the starting point for a new ontology of biobanked life in Europe.

### *Organ, tissue, fluid, and their harmonious recontextualisation*

How exactly new biobanks, samples, and information can be banked without bodies is a question that cuts to the heart of modern medicine. Organs and tissues have been studied and conserved far longer than their half-century history of clinical transplantation might suggest (Guibert et al. 2011; Watson and Dark 2012). However, while the organs and tissues have been (as they still are) an object of great interest for medical treatments and pathological diagnostics, they have become much-sought-after objects of informatics-heavy biomedical science, together with other biomedical samples. Fluids, tissues, and organs are important assets insofar as they are removed from bodies and recontextualised within a field of information (pertaining to the central characteristic features of the samples, the conditions of harvesting, location, etc.).

Biobanks are an innovation in the recontextualisation and regrouping of objects previously found within the body and limited by the skin. New tongues such as the lexicon presented earlier only create the conditions, the informational ecosystem where the decontextualised objects can be reinserted and revitalised. What is required, however, is precisely a 'harmonisation', or a harmonious body of information that couches the sample and its informatic representation. The MIABIS model is based on the premise of 'minimal information', the idea that there exists an informational content to all vital objects that every biomedical researcher (and, beyond the biomedical field, all researchers of the 'bio') can agree upon. But how does one agree on that minimal body of information to recontextualise the organs, tissues, and fluids?

When interviewed in August 2014, the individual responsible for MIABIS coordination recounted the group's way of tackling the problem:

> The logical first step to connect the biobanks is to know what they are and what they contain, and this is what we aim at here. I think we are successful because we were first with making a 'minimum dataset for biobanks' that could be adopted by all biobanks, around the world, by asking ourselves what the information is that all biobanks have in common.

The minimum, however, ended up being the minimal set of descriptive data on the donors, the samples, and the studies in which the biobank material is collected – a meta-ontologisation of the living bodies actually embodying all of this, in and as a sovereign person living within a sovereign country within the EU (see Figure 3.2).

Here we could extend the words of Foucault, who once identified the creative power of governance with the context of biodata. With the harmonisation (definition of the core dataset that describes biobanks and their samples), intensification of the data on the vital objects previously bound by individual bodies, with problematisation of health and its operational terms, ensued in the BBMRI project (cf. Foucault 1970, 122–123). This applied a logic of biopower and governance that pools and aggregates data on populations of choice, and here the population gathered is a novel collection of organs, fluids, tissues, and other vital samples from human beings. This is a governance logic typical of modern power, one wherein individuals are valuable insofar as they are recorded and can be seen and identified as part of a population-level aggregate. It also represents a new way of managing biobanks, wherein the aim no longer is to bank large numbers of people for one purpose, on population level and governed with one model. Rather, the individual is 'debunked' and the contents of that individual's body are laid bare, decontextualised, and ultimately recontextualised in new organ, sample, or fluid-sample populations. The logic of population is not based on the human, the sovereign, or the organism. It hinges on 'sample', at the same time smaller than an individual human being and larger than an entire human population.

As with the question of 'sex' in the Victorian era, addressed by Foucault (ibid.), the idea of linking biobanks found a new manner of expression, now in mutated form. Again, instead of directly using medical-registry data on patients and their samples, the biobanks started using aggregated data, thereby intensifying the data on life and health. These involve not one biobank, sample collection, or study but several at once. Today, MIABIS covers three core datasets, describing biobanks, sample collections, and studies, complete with thirty-seven attributes by which all of these are characterised in detail in a manner that enables implementation via a general standard for integration of Europe's biobank databases. This information model is centred on 'metadata' that include details of biobanks, the

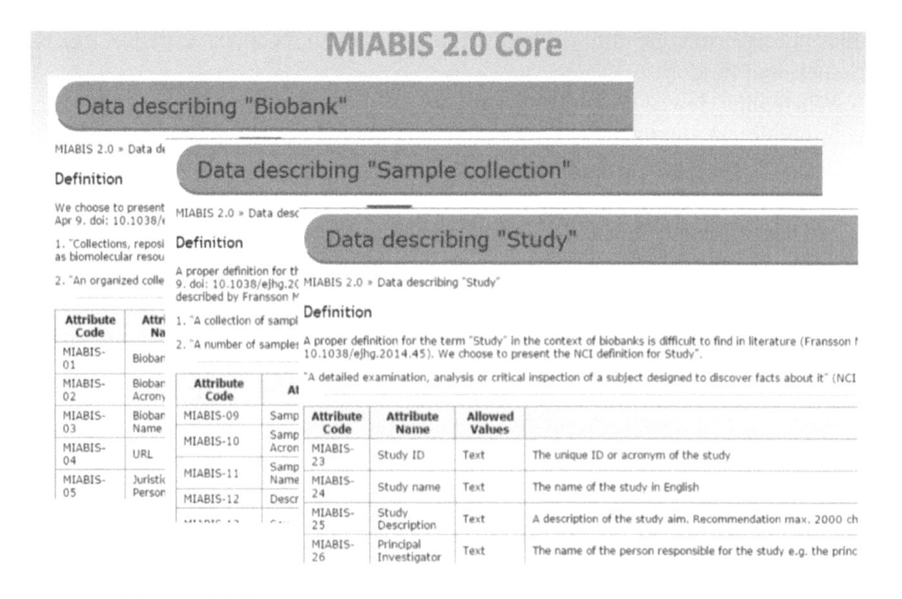


*Figure 3.2* MIABIS 2.0 core

sample collections they hold, and studies performed on the sample collections at an aggregate level. Thereby, the model escapes the restrictions of legal provisions for protection of personal data issued at national and European level. It thus provides a perfect, 'meta' bank for biodata.

The example of MIABIS shows nicely what is happening to our old ideas of wholes, actors, and sovereignty in the composition of tomorrow's life. The dominant model for human life no longer involves a person made of a head, torso, organs, and so on, and the same is true for the entire institutional way of imagining, putting together, and governing a nation in the globalised world. But, as does the database of biobanks, the data about European population show that there are no sovereign bodies at individual or population level. Organs do not compose sovereign bodies anymore; rather, populations and individuals as the objects of biopower are transformed into 'samples' that regroup to new populations because they are couched in new informational bodies that are standardised (e.g., under MIABIS standards). Hence, the metalanguage of today's life is not the state, the sovereign. Life today is articulated within the biobank, and population is articulated not internally to state boundaries but in HTML.

## Metacodes and technologies of the body

Biology in its essence is a technology itself bundled in bodies and separated from other bodies by the skin, or, as Robert Carlson puts it:

> Biology is the oldest technology. Throughout the history of life on Earth, organisms have made use of each other in sophisticated ways. The ancestors of both plants and animals co-opted free-living organisms that became the subcellular components now called chloroplasts and mitochondria. These bits of technology provide energy to their host cells and thereby underpin the majority of life on this planet.
>
> (Carlson 2010)

The redoing and reconfiguration of *anthropos* at will – the skin of mankind we all wear today and that we are due to shed in the future for an even better skin – is performed by metacodes that draw the line between the newly co-opted mankind of the twenty-first century and *Homo sapiens sapiens*. The face of the man once drawn in the sand of a quiet beach (Foucault 2002, 422) has been erased, but remaining fleshy materials and embodied information are put in service to the machinery as a sex organ for an ultimately connected new species. Our skin, tongue, and organs are all becoming fully connected, and by that development we are becoming the wealth and technology that we ourselves optimise through our bodies – as individuals, socially, and institutionally:

> Physiologically, man in the normal use of technology (or his variously extended body) is perpetually modified by it and in turn finds ever new ways of modifying his technology. Man becomes, as it were, the sex organs of the

> machine world, as the bee of the plant world, enabling it to fecundate and to evolve ever new forms. The machine world reciprocates man's love by expediting his wishes and desires, namely, in providing him with wealth.
>
> (McLuhan 1995, Chapter 4)

The wealth and problems generated by these kinds of machines, platforms of accumulation, and mechanisms of proliferation are what we turn to in the next chapter, where we consider them in the context of genetic resources beyond human species.

## Notes

1 This chapter is adapted from a previous version, which was published in: Tamminen, S. (2015). Bio-objectifying European Bodies: Standardisation of Biobanks in the Biobanking and Biomolecular Resources Research Infrastructure. *Life Sciences, Society and Policy*, 11, 13. http://doi.org/10.1186/s40504-015-0031-1
2 Physiologically, man in the normal use of technology (or his variously extended body) is perpetually modified by it and, in turn, finds ever new ways of modifying his technology (McLuhan 1995, 46).
3 There are multiple readings of this idea of 'error'. One is that of celebrated philosopher of the life sciences Georges Canguilhem, who developed a whole philosophy that secretly celebrated the creativity of life through error, in a vitalist approach. Another conceptualisation takes on error as a source of generative control, of biopower; here, the 'error' of our species is at the same time individualising and subjected to control by the modern state (Talcott 2014).
4 Examples include the Forum for International Biobanking Organizations, the International Agency for Research on Cancer, the International Society for Biological and Environmental Repositories, and the Public Population Project in Genomics (P3G).
5 For example, the European Biobanking and Biomolecular Resources Research Infrastructure, the Biomarkers Consortium, the Human Variome Project, and the UK DNA Banking Network.
6 For example, the Database of Genotypes and Phenotypes, the Encyclopedia of DNA Elements (ENCODE), the International HapMap Project, and the Open Biological and Biomedical Ontologies.

## References

Bateson, G. (1972) *Steps to an Ecology of Mind*. Chicago: University of Chicago Press.

Canguilhem, G. (1994) *A Vital Rationalist: Selected Writings From Georges Canguilhem*. New York, NY: Zone Books.

Carlson, R.H. (2010) *Biology Is Technology: The Promise, Peril, and New Business of Engineering Life*. Cambridge: Harvard University Press.

European Science Foundation. (2008) *Population Surveys and Biobanking*. Available at: www.esf.org/fileadmin/links/EMRC/SPB32Biobanking%5B1%5D.pdf (accessed on February 2015).

European Strategy Forum on Research Infrastructures. (2006) *European Roadmap for Research Infrastructures: Report 2006*. Luxembourg: Office for Official Publications of the European Communities.

Foucault, M. (1970) *The History of Sexuality, Volume 1: An Introduction*. London: Penguin.

Foucault, M. (2002) *The Order of Things: An Archaeology of the Human Sciences*. London: Routledge.
Guibert, E., Petrenko, A., Balaban, C., Somov, A., Rodriguez, J., and Fuller, B. (2011) Organ Preservation: Current Concepts and New Strategies for the Next Decade. *Transfusion Medicine and Hemotherapy: Offizielles Organ der Deutschen Gesellschaft fur Transfusionsmedizin und Immunhamatologie*, 38: 125–142.
Hadley, C. (2004) Biologists Think Bigger. *EMBO Reports*, 5(3): 236–238.
Harris, J.R., Burton, P., Metspalu, A., Palotie, A., Perola, M., van Ommen, G.J., and Zatloukal, K. (2012) Toward a Roadmap in Global Biobanking for Health. *European Journal of Human Genetics*, 20(11): 1105–1111, 1110.
Hayles, N.K. (1999) *How We Became Posthuman: Virtual Bodies in Cybernetics, Literature, and Informatics*. Chicago: University of Chicago Press.
Hutchins, E. (1995) *Cognition in the Wild*. Cambridge, MA: MIT Press.
Kay, L.E. (2000) *Who Wrote the Book of Life? A History of the Genetic Code*. Stanford, CA: Stanford University Press.
Lock, M. (1993) Cultivating the Body: Anthropology and Epistemologies of Bodily Practice and Knowledge. *Annual Review of Anthropology*, 22: 133–155.
McLuhan, M. (1995) *Understanding Media: The Extensions of Man*. London: Routledge.
Merino-Martinez, R., Norlin, L., van Enckevort, D., Anton, G., Schuffenhauer, S., Silander, K., Mook, L., Holub, P., Bild, R., Swertz, M., and Litton, J.-E. (2016) Toward Global Biobank Integration by Implementation of the Minimum Information About BIobank Data Sharing (MIABIS 2.0 Core). *Biopreservation and Biobanking*, 14(4): 298–306.
Mol, A. (2002) *The Body Multiple: Ontology in Medical Practice*. Durham, NC: Duke University Press.
Norlin, L., Fransson, M.N., et al. (2012) A Minimum Data Set for Sharing Biobank Samples, Information, and Data: MIABIS. *Biopreservation and Biobanking*, 10(4): 343–348.
Talcott, S. (2014) Errant Life, Molecular Biology, and Biopower: Canguilhem, Jacob, and Foucault. *History and Philosophy of the Life Sciences*, 36(2): 254–279.
Thacker, E. (2010) *After Life*. Chicago: University of Chicago Press.
Vermeulen, N., Parker, J.N., and Penders, B. (2013) Understanding Life Together: A Brief History of Collaboration in Biology. *Endeavour*, 37(3): 162–171.
Vermeulen, N., Tamminen, S., and Webster, A. (2012) *Bio-objects: Life in the 21st Century*. Farnham: Ashgate.
Watson, C.J. and Dark, J.H. (2012) Organ Transplantation: Historical Perspective and Current Practice. *British Journal of Anaesthesia*, 108: 29–42.

# 4 Crossing boundaries

## The global politics of access and plant as species of life™

### Introduction

A new generation of 'miracle crops' is being engineered and launched globally. Expectations could hardly be any higher: crops are needed to feed the poor, cure the ill, resolve the planet's biodiversity crisis, and improve the weather. Such expectations are the norm; increasingly, the language of global targets refers to billions of people and the promises span decades.

The most obvious places to locate such marketing are the polished advertisements of global agri-business. The product is showcased through its identification with agricultural landscapes and families eating together while the voice-over might tell us that a better tomorrow requires innovations that help farmers grow crops that meet 'the challenges of a hungry planet' (as in the rhetoric of Bayer crop science). Variations of this narrative point to the naturalistic foundations of innovation by recalling how 'from the very beginning, food is more than just a meal' (e.g., Monsanto) or pointing to the future as a chance to bring 'a new dawn' (e.g., DuPont). Such a future depends on 'collaboration between business and humanity' (Dow Chemicals) and will take the shape of a return to nature that depends on corporations that 'bring plant potential to life' (Syngenta).[1] These slogans are those of a sector wherein a few companies compete over the global markets for improved seeds and agricultural chemicals. Mergers and take-overs have been going on for decades, with the resulting conglomerates buying up plant breeding companies and each other. In 2015–2016, Syngenta rejected an offer by Monsanto to accept another offer from ChemChina, a state-owned chemical company. In turn, Monsanto rejected an offer from Bayer, while Dow and DuPont are merging. The amounts reported are never below the tens of billions, and this has been the case for decades in a dynamic that has been anticipated and well-documented since the late 1970s and early 1980s, revealing a pattern of corporate concentration and integration of commodity chains for crops, plant biotechnologies, and chemicals (Mooney 1979; Yoxen 1984; Goodman et al. 1987; Wield et al. 2010, 349).

What has changed, therefore, is not that such 'plant-branding' reflects the scale of the global markets involved but that the slogans match how innovation has become a privileged solution to inter-governmentally agreed sustainability goals, whether related to biodiversity, food, health, climate, or otherwise (Deibel et al. 2014). It is this species of 'plant-branding' that we set out to examine in this chapter. The idea is captured in the trademark sign ('TM') in the subtitle. On the one hand, the trademark points to the continuation of the commodification of plants, in a by-now-familiar process that should cause little surprise to anyone. Accordingly the trademark is one of many types of intellectual property that have been introduced to the life sciences in an attempt to guarantee a return on investments in crop science and plant biotechnology, along with stock trading, venture capital funding, asset management, etc. (Cooper 2008). On the other hand, the analysis in this chapter departs from the trademark as a symbol that refers to marketing and branding as integral to the mixing and hybridising of attempts to overcome the biological limits of plants in the life sciences and the notion that there might be ecological limits beyond which mankind should not go.

By now, it has become insufficient to simply document how it has become exceedingly complicated to establish who owns a plant and who decides about its usage. While this is true and lamentable, it is also ever more secondary as a concern to how the global environment of treaties and governance mechanisms has, over the last 15 years, combined the rules that enable the commodification of plants with voluntary commitments and non-binding targets as the preferred method for coordinated action on sustainable development and climate change. In fact, species of plants are exemplary in this regard. They are iconic of the commodification of life as a long-standing and heavily contested subject of global treaties alongside genetic engineering and related issues such as industrial agriculture and patenting, which have been debated since the 1980s.

In this regard, the title – 'Crossing boundaries' – refers to the transgression of biological limits in the life sciences, as well as in the specialised 'boundary work' that transforms plants into species of life™ (see Haraway 1997, 60; Halffman 2003; Gieryn 1983). Within this context, the subtitle invokes the language of access, as it performs a type of boundary work that facilitates continuous negotiation over the possibility of exceptions to the exclusive rights over biological resources that have been established as the norm for global environmental governance. Such a 'global politics of access' is then about how exceptions are being made to the commodification of life or, rather, how certain types of activities are eligible to be exempted from protection of intellectual property or the various sovereign claims to genetic resources (Safrin 2004).

Exceptions are being made to the established language of rights that establish state authority and ownership. Increasingly 'access' refers strictly to the usage of plant materials as genetic resources in the context of the life sciences and the development of related technologies. In such instances the language of access is invoked for arguing that a contribution is made to reaching global targets and fulfilling UN mandates. Considered together, however, the instances wherein exceptions are possible can be shown as, in fact, a norm; the language of access is a species of biopolitics. By establishing 'how' plant materials are being governed, we will argue that these are not simply exceptions – a marginal project of dubious effectiveness – and that they demonstrate a feature of a global 'state of exception'

(Agamben 2005). What is key is how advocacy of 'access' is turned into a means whereby 'legal determinations – and above all the very distinction between public and private – are deactivated' (ibid., 23).

The first section that follows elaborates on the discussion to make observations on the relationship between, on the one hand, the materiality of plant biodiversity and the embodied know-how about living and working with multi-species environments and, on the other hand, the juridical understandings of concepts such as commodity, enclosure, commons, and access. Continuing with this perspective, the subsequent section recontextualises and re-materialises the internationally negotiated texts and related governance mechanisms dealing with plant materials. The discussion addresses established topics such as 'access and benefit sharing' but also less well-known treaties and governance mechanisms wherein access is negotiated in regard of the usage of plant materials as technological resources, in initiatives, projects, and governance networks.

The conclusion reinterprets the analysis. Thinking across the examples in this chapter makes it possible to reconsider the commodity in a context wherein high-tech forms of gifting, charity, and altruism are privileged by how they are situated between law and life. The analysis shows that biological resources only to a limited extent are being incorporated into the body of the state – as its living embodiment, a foundation for its constitutional ordering of nature and the basis of the wealth of nations (see Tamminen and Brown 2011). While it is certainly true that the interplay of sovereignties is becoming more complex in consequence of the governing of biotechnological resources, the exception turns into the norm (through continuous negotiations of access) as congruent with biogenetic ways of thinking and acting. The key point, however, is to observe how the exercise of power comes to rely on and revolve around the 'opening up of a fictitious lacuna' (Agamben 2005, 31). It is through fictions that we can understand the consolidation of the detachment of global environmental governance from the materiality of the natural world as a 'zone of anomie' (ibid., 50). This zone is more and more exclusively inhabited by 'species of life™' that have become unabashed in imagining their own accomplishments through their association with global targets, as seen in UN sustainability partnerships and increasingly delirious attempts at plant-branding.

## Access and enclosure

### **Robinson Crusoe** ***as a myth of invention***

> 'the Concern I had been in for my own Preservation had taken off the Edge of my Invention for my own Conveniences; and I had dropp'd a good Design, which I had once bent my Thoughts too much upon; and that was, to try if I could not make some of my Barley into Malt, and then try to brew myself some Beer: This was a really whimsical Thought, and I reprov'd my self often for the Simplicity of it; for I presently saw there would be the want of several things necessary'.
>
> Daniel Defoe, *Robinson Crusoe* ([1719] 2007, 142)

The quote from Daniel Defoe's *Robinson Crusoe* is about a tragedy that is in sharp contrast to the familiar economic interpretations of the story of a man alone on an island. Such economics tend to omit how Crusoe spends most of his time and invests much of his labour in his 'designs', losing the edge of his invention when he becomes too preoccupied with defending his life and the possessions that he needs for his survival. This is a tragedy: innovation stopped once the island was accessed and he felt himself unsafe and compelled to work on his enclosures. Famously, Marx and Engels interpreted the story about a man stranded on an isolated island by pointing to his labour and the use-value of the things he makes, turning him into a model of pre-capitalist. This means that self-preservation – the principal natural right of life for Hobbes, Locke, and most other natural philosophers – is seen as lost in the context of the civilised world and the bourgeois economic system (Marx 1990 [1867], *Capital* Vol. 1 Section IV). Simultaneously, this kind of use-value consideration, ascribed to Crusoe on his island, should be seen solely in the context of his isolation, his isolation from the rapidly consolidating colonial world of empires. After all, Crusoe's labour, including his ability to invent new techniques and breed crops, belongs to the world of cash crops, spices, and many other valuable commodities that were already being distributed between continents (see Kloppenburg 2005; Parry 2004).

From this perspective, his inventiveness and his enclosures are part of an expansion of a global network that successfully enrolled new agents, artefacts, and distant peoples by interlinking dedicated spaces, such as plantations, model farms, and botanical gardens. This includes his independence, which should be understood as the mirror image of the exploitation of the labour of those who sooner or later found themselves caught up in the expansion of the European world. Hence, Crusoe on his island needs to be seen critically, as exemplifying the values of the British Empire and its projection of the power of colonial man over the natural world. The story, in this version, is exemplary of how white males are threatened by those who lived outside their societies, subjecting them into a labour force stripped of their natural rights as human beings (see, for example, Joyce [1912], cited by Barry, 2012, and Said, 1993).

Obviously, such a critique of colonialism remains valid; after all, Crusoe's story famously dehumanises the threat and violence of the savages who surround him by introducing us to Friday. Soon after the encounter, Friday puts his head on the ground in gratitude – 'with all the possible Signs of a humble thankful Disposition' – and Crusoe makes him 'know his Name should be Friday' in memory of the day that his life was saved. After a short while, Friday becomes like a 'child to a father' and wants to tell his fellow savages 'to live Good, tell them to pray God, tell them to eat Corn-bread, Cattle-flesh, Milk, no eat Man again' (Defoe 2007, 189). With Friday's conversion to a more civilised life, they proceed to protect themselves from those who would threaten them by enclosing more terrain and by beating and sifting corn. Soon after, Crusoe observes with approval that Friday:

> soon understood how to do it as well as I, especially after he had seen what the Meaning of it was, and that I was to make Bread of; for after that I let him

> see me make my Bread and bake it hot, and in a little Time Friday was able to do all the Work for me, as well as I could do it my self.
>
> (ibid., 179)

In sum, Friday is made to acknowledge the value of becoming 'civilised' so that he can be voluntarily put to work in support of Crusoe's designs. Our version, however, explores this tragedy by not only considering the narrow-minded manner wherein Crusoe puts Friday to work, but also pointing to how the events on the island foreshadow the liberal myth of the individual genius as the avatar of the global knowledge economy. We return to the figure of Robinson Crusoe precisely because he sits between the celebration of his work as a type of pre-capitalist use-value and innovation as the only recourse for the West in the age of globalisation. In this version of the story, the history of innovation is a tragedy, and Crusoe giving up on his inventions stands for our own predicament with industrial agriculture, beginning with Crusoe as a figure who belongs to the long history of plant breeding and the loss of crop diversity as indissociable from the rise of science-based plant breeding and plant biotechnology.

### *Access, insecurity, and enclosure*

There has been a constant demand for and necessity of resistances to disease and adapting to changing soil and weather conditions ever since the beginning of agriculture. Crops are innovations that are the result of how seeds are selected and saved by farmers every harvest, reproducing only those plant varieties that show the most desirable traits. Notably, Crusoe is one of them, a breeder-brewer-inventor who represents the historical relationship between the distribution of crop diversity across the world, especially in the colonial age, and the ongoing accumulation of varieties and traits. Together, these processes resulted in the variety of food crops that were adapted to and that shaped the diversity of human practises and ecologies, in forests, wetlands, pastoral lands, irrigation systems, or otherwise. However, this wealth did not last, just as Crusoe had to give up on his inventions when no longer in isolation. The book tells us of Crusoe the breeder/brewer/inventor who labours until he finds a footstep in the sand. He had survived the decades left to himself after his shipwreck by getting involved in breeding and brewing with the barley he had salvaged. He stops inventing, however, when recognising that there are others: he repeats several times that it is 'on the account of the Print of a Man's foot' that he leaves behind his earlier designs (ibid., 130–132). He starts planting rings of trees in hedges that become so thick and strong that they are as walls of fortifications wherein he could hide himself and enclose himself, his crops and the cattle that are his immediate supply of milk and meat. He admits to himself that this 'Wall I was many a weary Month a finishing, and yet I never thought my self safe till it was done' (ibid., 136, 137). After he finds the footprint in the sand, he asks himself: '[what] if it should happen so that they should not find me, yet they would find my Enclosure, destroy all my Corn, carry away all my Flock of tame Goats, and I should perish at last for meer want' (ibid., 132).

There are three key moments from Crusoe's tragic history of innovation that are the starting point for turning to our own predicament: the rapid extinction of native crops, local cultivars, landraces, and wild relatives that are cultivated in more variable environmental conditions. These are the moments of (1) access (to his island by others), (2) insecurity (about his ability to survive), and (3) enclosure (labouring to protect himself and his possessions). In the remainder of the chapter, we will examine the interplay of these three terms in a triangle of pivotal moments that can be recontextualised to understanding the political dynamic of how plants are being transformed into 'species of life™'.

### *Seeds and the island-empire*

The most iconic example of 'plants as a species of life™' is probably the bag of seeds that is sold with several conditions attached. These conditions are possible because of the patenting of DNA and have increased the scope for capturing profits; for example, there might be a type of insecticide that becomes mandatory, or it may be illegal to replant the variety after its cultivation or use its seeds for breeding. The consequence has been corporate concentration and intensification of the transformation of farmers into customers.

How can we perceive such commodification in terms of how Crusoe ended up enclosing his possessions? Crusoe stopped innovating even though he remained convinced of the potential of his innovations – were it not for 'the terror' he was in, he would have 'undertaken it, and perhaps brought it to pass too' (ibid., 142). Instead, he feels unsafe after finding the footprint, which resembles how the free flow of plant materials has been undermined. Jack Kloppenburg, in his seminal 'First the Seed', a classic history of plant improvement, uses the metaphor of an 'island-empire' to explain how the agricultural sector historically lay outside the 'capitalist mainland' (Kloppenburg 2005, 280). Kloppenburg discusses the new plant biotechnologies as following a historical trajectory that mirrors and intensifies the earlier focus on crop sciences, which in the twentieth century had opened up 'a whole new frontier of accumulation' and constituted a 'breach' of the biological barrier that had prevented more than a bare minimum of private investment in crop improvement (ibid., 11). This, then, is the footstep, the moment of access, that is followed by what he calls 'the construction of bridges', intensifying of access through the subsequent development of plant biotechnologies. After the island has been opened to outsiders, bridges begin to be built that would permanently connect the island to the capitalist mainland (ibid., 279). What is crucial is that enclosure follows access. On the one hand, constructing enclosures is conditioned on the potential for opening up the 'island-empire' to investments from the mainland. On the other hand, such access needs to be consolidated in regard of the remainder of the island-empire, which represents untapped potential, a commonwealth to be commodified. Accordingly, it is through the transformation of plants into species of life™ that access is consolidated, that the remaining limits are to be overcome. Plant-branding is most significant in a context wherein there are limits to the access that has already been established: when bridges have been built, there remain obstacles and resistances that undermine the capturing of value and the message that it is just a matter of time before, eventually, the expectations are met and science will deliver on its promises about resistance to disease, tolerance of any kind of weather or soil conditions, and whatever other biochemical properties might be in demand on the world market.

In other words, agency is identified with technological change, driving the integration of plant materials into the commodity system and establishing as incontestable that new crops are of greater value than dynamic usage of breeding resources and locally specific improvement trajectories that are still around. The latter need to be devalued if the potential for profits is to be realised and if access to the common-wealth is to be kept secure. For example, such devaluation can be seen in how the complexity of crops is presented as a mere setback, deliberately underestimated in the discussions of biotechnological change. Similarly, the rules and regulations establish that there should be no meaningful competition, economically or morally, with farmers who might choose to continue as breeders of their own plant varieties and as competitors to the commodification of plant varieties. As Crusoe did, many choose enclosure when faced with the insecurity that arises with a footstep, and with the subsequent constructing of bridges from the mainland. The point here is not that science-based crops essentialistically threaten an Arcadia that is unexplored and unknown, where the diversity of plant life exists beyond ownership. Instead, it should be clear that only certain types of agricultural commodities are being traded globally and are valued through eco-financial speculation, while other types of crop cultivation are continuously devalued in combination with their relations to the livelihood of millions or even billions of people.

## Plants and global treaties

### *From access and commons to enclosure*

If science-based breeding is the footstep, than the best candidate for Kloppenburg's 'bridges from the mainland' is the patent. Just as Crusoe came to fear for his possessions, the patent is the mechanism that represents how new claims to the biogenetic wealth of the island-empire could be made, to be countered in various ways by securing of its resources, either economically or as integral to the well-being of society and its relationship with nature.

Such a response is visible in the failed attempt to declare that plant materials are common property or, rather, a 'common heritage of mankind'. Patents on DNA were authorised in the early 1980s, which explains why the opening statement of the Food and Agricultural Organisation's (FAO) 1983 International Undertaking on Plant Genetic Resources announced that 'plant genetic resources are a heritage of mankind and consequently should be available without restriction'.[2] This treaty was not a denouncement of patents; it was primarily an attempt to recognise the moral or material value and utility of the labour that farmers have been expending in the development and regeneration of crop genetic diversity.

To be specific, farmers do not receive any benefits when the varieties they produce are taken up for commercial use, yet they are expected to purchase the varieties developed by corporations as exclusively owned products at higher prices. There has never been and is no reciprocity between farmers and the seed companies that are obtaining and using the shared crop genetic resources, receiving a surcharge on the sale of proprietary varieties. Crops are a natural resource that cannot be depleted through usage, while their survival depends on open access and non-exclusion as the norm in farmers' fields (Brush 2005a, 2005b). Accordingly, the loss of crop varieties is an excellent example of failure to govern the commons, as any varieties that are not used by farmers go extinct. The rate of extinction of crop biodiversity in the twentieth century is directly related to the use of science-based varieties in industrial agriculture (Gepts 2004; Hughes and Deibel 2006; van Dooren 2007, 2009).

Farmers always look for new traits in response to diseases or act on changing conditions, which implies that seeds with characteristics that are in demand are more valuable. However, this dynamic of demand, value, and scarcity of resources is very different from what we find when considering the same concepts in the context of the introduction of new intellectual property incentives. There are not only patents, but also 'plant variety rights' and 'plant breeders' rights' were already introduced to conventional plant breeding in the mid-twentieth century. The conflict between the two types of economics has as its result a 'paradox of plant breeding' that has to do with the creation of new varieties by crop scientists and how these have 'historically undermined the very genetic basis on which [crop science] rests' (Gepts 2006, 2281; Deibel 2013).

This paradox was the premise for designating plant materials as a common heritage of mankind, which affirms that diversity is sustained through the usage of crops. This was a remarkable episode in international law because other examples that are sometimes seen as a common heritage involve resources that were freely available only as long as they could not be claimed in any practical sense. Enclosure is impossible until it is within reach, which is to say that claims begin to be made after polar expeditions start going to Antarctica, when satellites go into space, when submarines start exploring the deep sea-bed, and so forth (see Shackelford 2009). The latter are commons without commoners, neither the irresponsible types personified by Garrett Hardin's herdsmen who overuse their own commons nor those who know better than to plunder the resources on which their livelihood depends. In Hardin's 'tragedy of the commons', access to the shared resource results in its destruction (see Hardin 1968). A popular response to this type of tragedy is to observe that common property is often the most sensible alternative to the failures of privatisation and state-led forms of policy making. Such 'governance of the commons', a term coined by Elinor Ostrom, who received a Nobel Prize in Economics for it, is regularly used in referring to communities' preservation of plant biodiversity as an example of successfully managing a common-pool resource by strengthening local decision-making processes (Ostrom 1990, 2005).

The problem, however, is that by showing the limits of Hardin's view, we find that there are many co-existing types of common property, without a clear perspective on how we inevitably end up in a situation wherein one type 'may need to be protected at the expense of another' (Harvey 2011, 102). Giving priority to certain types of commons rather than others might not be what commons theory intends at all. Nevertheless, this is the implication of how exactly crop biodiversity is being considered as a commons that is threatened with enclosure. Should crop biodiversity as a common-pool resource be prioritised, or is it secondary to the 'knowledge commons', which has increasingly gained traction as a theory of innovation? This refers to 'the tragedy of the anticommons', in which innovation slows down when too many parties hold exclusive rights to access knowledge or technological resources. The result is that the usage of technologies, resources, and the production of knowledge is obstructed and innovation is frustrated (Heller 1998; Heller and Eisenberg 1998). Originally, the argument about the anticommons was about how the patenting of DNA is discouraging innovation in the life sciences, which implies a critique of patents and a defence of how many technologies are managed as common-pool resources. Conflict ensues when plant materials are valued primarily as a technological resource that needs to be shared to encourage innovation in the life sciences.

The question is, therefore, 'how' genetic resources need to be shared. This includes 'how' to set the criteria for successfully managing common-pool resources but also 'who' needs to be prioritised as the commoner. In that regard, it is key to begin by recognising that crops have always belonged already to those who used them, long before plant materials came to be considered as technological resources within the context of sophisticated techniques for the breeding of plant varieties. Their usage was vital to the creation of plant biodiversity that survives within a context of distributed and communal activities that display the regulated character of 'continuous sharing of many physical resources on informal rules and customs that are developed and adapted over long periods of time' (Drahos 1996, 12; Hess and Ostrom 2003).

Yet this is not what happened when DNA was declared to be a technical domain through the introduction of patents in the 1980s. Not only were temporary monopolies over inventions authorised in the life sciences; its implication was that the DNA was no longer legally a 'manifestation of nature'. DNA would henceforth be a 'product of ingenuity' and a 'composition of matter' (see Parry 2004, 85, 86; see Carolan 2010; Calvert and Joly 2011). This is than the response to finding the footprint: the moment when exclusively owning DNA became possible coincided with the transformation of how the usage of plant materials was valued. The future of the diversity of plant life, from that moment on, would be seen primarily through a lens that shows the potential return on investments in the life sciences. The decades that followed have seen such eco-financial speculation become 'a highly profitable – indeed rational – enterprise' (Cooper 2008, 24–28). The remainder of the discussion will demonstrate the implications of what is meant with eco-financial speculation, describing a socio-technical setting wherein the accumulation of biological futures is normalised as the step that succeeds the ability to enclose the 'very principle of generation, (re-)production itself, in all its emergent possibilities' (ibid.).

We observe the same logic when examining 'global environmental governance', which increasingly deals with farming and biodiversity in terms of the establishment of knowledge commons as the model for the nexus of human practises and ecologies, whether as seeds, forests, wetlands, pastoral lands, and irrigation systems or otherwise.

### *Access and sovereignties*

The international undertaking of the early 1980s was unsuccessful; it marked the only international treaty to have made the suggestion that plant materials should be free of private property and state sovereignty. Now there is the International Treaty on Plant Genetic Resources for Food and Agriculture (henceforth the Plant Treaty) from 2004, which is itself remarkable in that it is based on the recognition that access to plant materials is a benefit that depends on being shared.

The notion that plant materials were to be 'free to all men and reserved exclusively to none', which was put forth briefly in the 1980s, was removed quickly as a result of a provision that was added to the next treaty of the FAO, the Undertaking of 1991. This text highlighted that there is no incompatibility between plant breeding and the conservation of crop diversity, and the Convention on Biological Diversity (CBD) was anticipated by making explicit that it is within government's jurisdiction to restrict the free exchange of materials in order to conform to national and international obligations (Brush 2005a, 86). By the early 1990s, there was no longer a mandate for the FAO to identify the development of high-performance crops and their commodification with the loss of plant biodiversity. Furthermore, the CBD, which was ratified in 1992, did not make any special reference to plant materials in agriculture, while agriculture became part of the mandate of the World Trade Organization (WTO) in 1993 in parallel to Trade-Related aspects of Intellectual Property Rights (TRIPs) agreement (May and Sell 2006).

The first article of the CBD calls for the fair and equitable sharing of the benefits arising out of the utilisation of genetic resources (art 1. CBD). On the one hand, the CBD is a framework convention for several international treaties of relevance to species and ecosystems conservation. Its negotiation took ten years, and the end-result was a treaty that has its premise that quick financial returns on the destruction of natural resources need to be replaced with economic incentives that allow governments to focus on biodiversity (see Boisvert and Vivien 2012). On the other hand, the references to market- based conservation are explicitly imagined as a response to the development of plant biotechnology. This is shown by how the text focuses on 'genetic resources' with 'actual or potential value' as the basic material of the life sciences (Article 2 of the CBD; see Hayden 2003a, 2003b; Gepts 2004, 2006; Hamilton 2006).

Accordingly, many governments have interpreted the CBD as assigning them the responsibility for biodiversity found inside their territory and the authority to create regulatory frameworks to prevent the unauthorised and uncompensated exploration of biodiversity as commercially valuable genetic and biochemical resources (Reid et al. 1993). This is called 'bio-prospecting', and many countries have taken measures to address what they see as shortcomings in intellectual property rules in areas such as unauthorised access, insufficiently strict review of patent applications, and the duty to share benefits with stakeholders (Hoare and Tarasofsky 2007). International negotiations over 'access and benefit-sharing' have continued within the context of the RIO+20 framework, within the World Intellectual Property Organization (WIPO), and at the request of the FAO. The discussion has become increasingly technical, as the focus primarily lies on the obligations for intellectual property holds and the benefits that are to be given to specific groups without any legal rights being granted to those who live and work in biodiversity-rich areas (Lesser et al. 2007).

This includes the ratification of the Plant Treaty in 2004. It is unique as an environmental treaty that has been ratified while taking the position that access is the main benefit to be shared. Accordingly, its preamble states that access is vital for an adequate response to concerns over food security, the conservation of agricultural biodiversity, the sustainability of farming, and so on. However, the kind of access that can be arranged in the Plant Treaty still must be compatible with TRIPs and the CBD.

Following the ratification of the CBD, numerous national representatives began arguing that access to agricultural biodiversity could be restricted nationally, which meant that negotiations had to be started on a new multilateral system for plant materials in agriculture (Coupe and Lewin 2007). One of the results is that the arrangement of access refers to the content of seed banks under the authority of participating governments, the FAO, and the Consultative Group on International Agricultural Research (CGIAR). Hence, there are no obligations imposed on companies and governments to remove any restrictions on the usage of plant materials by farmers.

The FAO continues to call on governments to respect farmers' rights (to the saving, selecting, and replanting of seeds) but no longer considers these to be 'vested in the International Community, as trustee for present and future generations of farmers' (FAO Res. 5/89). The Plant Treaty has changed the landscape: farmers' rights are now something that governments might or might not act on. Furthermore, much of the discussion on access and benefit sharing is aimed at generating revenues for the FAO on the basis of a small percentage of the commercial value of finished plant varieties that restrict the accessibility of the crop varieties stored in seed banks. In other words, access is not guaranteed but taxed. The FAO collects revenues when plant materials in the system of seed banks become unavailable for research or breeding (ibid.; Gepts 2004; Safrin 2004).

## *The transformation of global governance*

Before we continue with the analysis of the global politics of access, it is crucial to make an observation on the transformation of the wider setting for global environmental governance wherein the response to the ratification of the CBD and TRIPs took shape.

As explained, the legal framework for market-based conservation has set in motion the corrosive interplay between the patent-based and the sovereignty-based system of ownership (Safrin 2004, 642). Expectations were raised about the potential to redistribute some of the value of genetic resources, and the negotiations over access and benefit sharing have dissociated concerns over the loss of plant life from market failures or excesses. Simultaneously, environmental governance became increasingly voluntary and non-binding through the UN's endorsement of the formation of partnerships. These are flexible governance mechanisms among various state and non-state stakeholders that are intended for bridging the public–private dichotomy, and represent a co-operative form of organisation wherein the rationales of various sectors can be applied efficiently (Bäckstrand et al. 2010; Andersen 2008).

Hundreds of partnership agreements have come into effect, and they cut across various levels of governance of nearly every issue that was on the environmental agenda after the 2002 World Summit on Sustainable Development (WSSD). At that time, sustainability partnerships were declared 'the way forward' (as 'type-II outcomes of the summit'), and many commentators in fields tied to plant materials echo the optimism surrounding their formation by adopting policy- oriented language of 'best practices', 'lessons to be learned', and 'success factors' (e.g., Spielman et al. 2010; Ferroni and Castle 2011; van Berloo et al. 2008). Many others, however, have doubts about the effectiveness of the 'partnership regime' at global levels and vis-à-vis the promised goals (Biermann et al. 2007). Accordingly there has been a general lack of success of the 'partnership regime' in regard of the various goals and the need to address governance deficits (Pattberg et al. 2012).

The formation of a UN partnership needs to be sufficiently agreeable to all parties involved, and it needs to serve the interests of all stakeholders, who are left free to identify their activities with the realisation of sustainability goals as they wish. This is because the WSSD did not result in a partnership regime that includes standards on the relevancy of stakeholders (businesses and NGOs), what the aim of partnerships should be (implementation versus participation), or how they should be screened and monitored. There are many examples; UN partnerships operate as platforms for controversial technologies being rebranded as sustainability projects, among them nuclear energy, PVC products, and water-purification chemicals and so on (Mert and Chan 2012). Other examples are less controversial but highlight how corporations are assumed to have a neutral if not benevolent role in respect of the realisation of sustainability goals. Examples include Dow Chemical sponsoring of the Blue Planet Run to 'bring safe drinking water to 1.2 billion people'. The Coca-Cola Foundation and Procter and Gamble promote the provision of a water disinfectant and 'behavior change techniques directed towards improved hygiene' in water-deprived poor countries, while Royal Dutch Shell is involved in the Clean Air Initiative to enhance air quality and reduce emissions.[3]

Similarly, there is no need for Monsanto's actions to be endorsed in international texts when it gets involved in a partnership that attempts to contribute to mitigation of 'the threat posed by invasive species' (UN CSD 2008). It is not needed for a UN partnership's mission to be endorsed in international texts, which would in this instance clearly show tension between Monsanto's business model and the Cartagena Protocol on Biosafety wherein the protection of biological diversity requires that measures are taken to control the risks of biotechnology. Furthermore, this example is an indication of the kind of plant branding reflected in the countless adverts, websites, and advertisements promising that the Great Outdoors might be saved when companies pledge to donate a percentage of the cost of flying across a green-blue planet, driving an SUV through a pristine forest, or eating a Happy Meal with ingredients that depend on soy-based animal feed or other elements of industrialised agriculture (see Büscher 2012). Environmental governance has now become fully compatible with plant-branding and actively engages with efforts to raise expectations about the economic and social value of corporations' products and services.

The point is not that there were no contributions being made by individual partnerships to achievement of the Millennium Development Goals (MDGs) or any of the goals set after these expired. Rather, the type of access to plant materials that is possible reflects a logic of environmental governance wherein conflicting and controversial practises can be acted upon without formal endorsement at the UN level (see Biermann et al. 2007; Mert and Chan 2012). This is essential context considering that the Plant Treaty was ratified at around the same time as the endorsement of partnerships. Many commentators have noted how this has led to UN organisations partnering with corporations that are driving controversial fields such as genetic modification, patents, and industrial production of food (Hisano 2005; Kaan and Liese 2010; Hatanaka 2009; Poulton 2012).

Furthermore, the restrictions that were put in place during the 1980s and 1990s, as in TRIPs and the CBD, are naturalised through the creation of closely related governance mechanisms that settle the logics (practises, actions, and language) derived from the previously negotiated texts (see Mert 2009). Its implication is that international organisations were transformed into partners in governance networks wherein they have to persuade others to remain or become embedded, relevant, and central to policy making. Such a logic of environmental governance continues to take its shape in response to developments in the life sciences. Access to knowledge (abbreviated as A2K) is widespread in the form of the cross-licensing of patents, the creation of joint ventures, private consortia, and partnerships of various kinds.

On the one hand, the advocacy of access is often described in terms of knowledge commons, or intellectual commons revolving around the successful management of common-pool resources (see Burk 2002; Boettiger and Burk 2004; Opderbeck 2004; Allarakhia and Wensley 2007; Overwalle 2009; Hope 2008, 2009). This includes occasional references to enclosure in an historical analogy to Marx's description of the enclosure of the English commons (see Harvey 2011), for instance in James Boyle's 'second enclosure movement' and Christopher May's 'new enclosures' (May 2000; Boyle 2003a, 2003b, 2008). Such critiques, however, tend to celebrate the commons in contrast to enclosure and in isolation (as ontologically distinct) from the intricate institutional balancing act of access and restrictions that is re-orienting global environmental governance to refer to very different sets of activities. On the other hand, our contrapuntal reading of the history of plant breeding gives us a position from which we can challenge and infiltrate the premise of a common world of shared, compatible, and global targets (more food, more health, etc.). In what follows, we discuss those governance mechanisms that show how the arrangement of access privileges follows a technoscientific rationale wherein experts are mobilised to work on strategically isolated exceptions to restrictions as a general rule for everyone else. What this demonstrates is the suspension of the relations of the juridical order of sovereignty as authorised through global treaties. Significant is 'how' this suspension is acted on, taking shape in response to the complex legal relations that have accompanied how global markets and technologies transformed the usage of plant materials.

### *Access to genetic resources*

One of the most prominent examples of market-based conservation in the time following the ratification of the CBD was the Diversa corporation, which was actively trying to establish a business model aimed at bio-prospecting. Crucially, however, the company no longer exists, and its successor operates on a very different idea of what genetic resources are, what they are for, and where it is in the life sciences 'that value resides' (see Helmreich 2008).

Firstly, Diversa's website used to refer to its activities as revolving around the accessing of

> a wide variety of extreme ecosystems, such as volcanoes, rain forests, and deep sea hydrothermal vents, to collect small samples from the environment to uncover novel enzymes produced by the microbes that dwell there. Because the harsh temperature and pH conditions in which these 'extremophiles' live often mimic those found in today's industrial processes, they are a rich source of potential products. Through the use of proprietary and patented technologies, Diversa extracts microbial DNA directly from collected samples to avoid the slow and often impossible task of trying to grow the microbes in a laboratory. We mine this huge collection of microbial genes, numbering in the billions, using ultra high-throughput screening technologies with the goal of discovering unique enzymes.
>
> (www.diversa.com/ as accessed in May 2007; off-line July 2007; see Deibel 2009 and also Helmreich 2009, 142)

On the one hand, this statement is very comparable with the CBD in how it highlights the value of biodiversity as raw material for industrial processes and potential products. The company specialised in accessing ecological and ethnobotanical knowledge that can focus or re-direct the screening of enzymes. To that end, the extremes of biodiversity have to be screened, sequenced, and turned into databases; profits are made once the information is re-materialised as food, energy, medicine, or other valuable commodities. On the other hand, Diversa's website began automatically redirecting to 'Verenium: the nature of energy™' in July 2007. The new site explains their aim is to make 'biofuels' 'from low-cost, abundant biomass' and create speciality enzyme products.

The name change is significant; no longer is any relationship being suggested to bio-prospecting. The website refers to 'new, greener methods to produce the fuels and other industrial products that are the lifeblood' and 'the building blocks' of 'our 21st century way of life'. What this re-phrasal of the more familiar 'black gold' suggests is that the disappearance of the Diversa corporation has to do with a devaluation of the 'green gold' in the life sciences, which suddenly is turned into an 'abundance' of biomass. In other words, profiting from bio-prospecting turned out to be more complicated than Diversa's mission statement suggested. The screening of samples and the commercialisation were neither easy nor quick, which is reflected in how many other pioneers of the field went out of business (e.g., California-based Shaman Pharmaceuticals) and in reports that suggest that the larger companies allocate a much more limited amount of their research funding to prospecting (Boisvert and Vivien 2012; Firn 2003).

Most reviews of bio-prospecting foreground the complexity of the contractual relations. Access to volcanoes, rain forests, the deep sea, and other extreme ecosystems are examples of the type of locations where numerous parties are involved and where it is going to be politically sensitive to determine and legitimise which communities to involve in the deal. For example, Stephan Helmreich has described at length how native Hawaiians" notions of ownership and innovation conflicted with Diversa's approach and how the former objected to its bio-prospecting activities (Helmreich 2009, 110–130). Furthermore, this case mirrors the most iconic and controversial instance of bio-prospecting, that of the well-known deal in 1991 between Merck & Co. and INBio, a not-for-profit private organisation (the Instituto Nacional de Biodiversidad) under the control of the Ministry of Natural Resources, Energy and Mines of Costa Rica. This NGO managed a park in Costa Rica where Merck hoped to identify natural compounds with interesting gene candidates for pharmaceuticals.

Notably, the INBio deal involved a substantial amount of money and resources, which to some has qualified it as a success, suggesting potential for more direct and decentralised models with flexible norms as a 'value-added approach' to market-based conservation (Safrin 2004). Others, however, have pointed out that the payments went towards the training and equipment necessary 'to access, identify, classify, and collect biological materials on Merck's behalf' and that it is 'difficult to underestimate' the significance of this example for the implementation of the CBD as it took 'on the force of fact on the basis of little more than hypnotic reiteration' (Parry 2004, 120–122, 216, see also Brush 2005b, 19–20; Boisvert and Vivien 2012). Here, however, the lack of convincing examples of bio-prospecting and its demise as a business model is more relevant. It demonstrates the difficulty of maintaining sovereignty as the norm in relation to the potential to derive wealth from genetic resources. To suspend the norm is, then, not a violation but a means to uphold the complex interplay of sovereignties in its relationship to the transformation of the usage of DNA (see Agamben 2005, 23). While 'access and benefit-sharing' is likely to retain broad support in international legal negotiations, the naturalisation of intellectual property can be followed from market-based conservation to the arrangement of access in the life sciences as integral to the process wherein the information extracted from living materials is rematerialised as food, health, energy, and so on.

### *Access to crop diversity*

The Global Crop Diversity Trust[4] has taken on the nickname 'the Arctic Vault' because its seed depository is located close to the North Pole. Since late 2007, the vault has ensured the conservation and availability of millions of seed samples from around the world that are copies of the samples from the major crop-diversity collections of gene banks all over the world. It was established by the FAO and the CGIAR shortly after the ratification of the Plant Treaty. Financial contributions came from government agencies, international organisations, multinational foundations, and corporations (including Syngenta, DuPont, and the Bill and Melinda Gates Foundation).

The coverage in the popular press at the time of its foundation was frequently Biblically evocative or otherwise measuring up to that scale, calling it the 'doomsday vault' and the 'ark of Noah'. The President of the European Commission at that time called it a 'frozen garden of Eden', and Carry Fowler, its director, termed it an insurance policy for the world's food supply, while critics have called it a 'giant icebox' with 4,5 million seeds.[5] Indeed, one of the reasons for the location was that it guarantees a temperature of -2 degrees Celsius due to the surrounding permafrost, even without electricity. Furthermore, this storage facility has as one of its objectives to act as a back-up for seeds and crops lost in natural disasters such as tsunamis and hurricanes or man-made disasters such as the wars that have destroyed gene banks and crops in Sierra Leone, Iraq, Afghanistan, and Syria. More practically, it insures the conservation of seeds against mismanagement and underfunding by governments that do not make crop genetic diversity a priority.

Accessing the collection closely mirrors the terms of the Plant Treaty as a multilateral system wherein access is the benefit to be shared. The FAO and the CGIAR can refer to the Arctic Vault as an example of its implementation, and it is often mentioned that everyone can access the collection – whether farmers, breeders, seed companies, or biotech multinationals. Indeed, the compensation scheme of the treaty contributes to the funding of the Arctic Vault. However, Genetic Resources Action International (GRAIN) argued immediately that the logic of the vault is to 'remove crop genetic diversity from farming: as people's traditional varieties get replaced by newer ones from research labs – seeds that are supposed to provide higher yields to feed a growing population – the old ones have to be put away as 'raw material' for future plant breeding'.[6]

The point is valid in the sense that a displaced community will not benefit any longer from their seeds having been frozen, as they no longer farm under conditions matching those local varieties. Also in other circumstances, the utility of re-introducing seeds hinges on the knowledge and skills of the farmers who use the variety. Once local varieties are no longer being used, their reintroduction becomes unlikely. To re-introduce rare varieties requires communities that have safeguarded the adaptive capacity of their food systems and kept working with crops as 'embodied know-how' of generations of farmers as a composite of 'associations, practices and ecologies' (Van Dooren 2007, 85). In this regard, it is much more likely that crops are lost because of the introduction of monocrops with traits that are valuable on the world market, which means that their storage no longer serves as a back-up for local seed usage. For example, climate change is one of the disasters that is mentioned, as it is expected to lead to the loss of crop biodiversity and also a decline in agricultural production. The vault is presented as a climate change safe house for genetic diversity that is integral to the development of 'climate-ready' crops for the areas that are most threatened, on behalf of a 'world [that] is changing faster than seeds can adapt in nature'.[7] Hence, a discursive relationship is established between the loss of crops' diversity and a wider sense of global insecurity. On the one hand, the Arctic Vault mirrors the Plant Treaty with a logic of governance wherein the rationale is that access is the main benefit to be shared, on the mistaken and persuasive belief that if a resource is open to all, it will be equally exploited by all (Sunder 2007, 106; see also Chander and Sunder 2004; Boyle 2003a, 2003b). On the other hand, the vault is remarkable in its affirmation of a dystopian future – unique as a global governance mechanism that prepares for insecurity rather than aiming to reach global targets. Accordingly, a limited type of access is universalised in how it suspends the exclusive claims (of sovereigns over their territory's genetic wealth as well as intellectual property protections). In other words, the norm is affirmed by the exception: enclosure continues, and no restrictions on the usage of seeds by farmers are removed. Furthermore, the exception to the norm is an indication of how innovation is privileged in environmental governance. No longer is it simply the case that expectations are made in regard of the commodity. The commercial value of crops is extended to the usage of plant materials everywhere else. Enclosure no longer is only about the rush to patent the genetic traits of crops when access refers to genetic diversity as a treasure hoard that requires maximum security at the ends of the earth against a disaster that is to occur in a near or distant (biological) future.

### *Access and innovation*

Universalising access for exceptional cases and circumstances occurs often with the UN's approach to sustainable development. This includes the international system for the management of plant materials, and the examples are indicative of a decline in plant materials as a common heritage. The aim is to arrange access to high-performance crops as a form of global altruism, and a new kind of 'romance of the commons' is affirmed through the adoption of humanitarian licences whereby techniques to work on high-performance seeds are donated to scientists.

The field of biofortification is one of the strategic areas of investment of the CGIAR and has gained prominence in recent years for its potential solution to malnutrition. The most well-known example comes from the late 1990s, when Golden Rice was launched and attracted great publicity because of the genetic modification of the traits of rice to express higher levels of vitamin A. Deficiencies in vitamin A cause blindness, especially among the poor in East Asia. The technology is still under development (e.g., it expresses insufficient quantities of the relevant substance, beta-carotene), and the work is now coordinated by the International Rice Research Institute (IRRI), one of the CGIAR centres. The project is one of many wherein plant improvements are made available using a 'humanitarian licence' guaranteeing that the technologies are available to 'subsistence users' around the world and to researchers on subsistence crops who cannot afford to licence biotechnologies.

On the one hand, the Golden Rice project shows how genetic engineering is being identified with environmental governance. As mentioned, this is not formally endorsed by the UN, and the CGIAR does not have a clear policy on plant biotechnologies. Traditionally, there are strong reservations about molecular biotechnology within the UN system and among the international representatives on the CGIAR board, and the policy of the CGIAR centres had been that germplasm, technologies, and research results should be available in the public domain. Golden Rice, however, was invented by scientists working for public institutions who struck a deal with AstraZeneca – now Syngenta – which holds the commercial rights. Golden Rice technology was covered by too many patents for its further development; reportedly, it involved some seventy process and product patents, held by thirty-two companies and universities (Kryder et al. 2000). A partnership agreement was negotiated wherein the companies guaranteed the distribution of the rice on a royalty-free basis to marginal farmers who live in developing countries, as well as sublicensing to public research institutions. The CGIAR licence mentions that a technology may be used, made, and sold without royalties for trade or business that results in monetary income of less than €10,000 a year per business.

Golden Rice set a precedent. For example, the HarvestPlus project (like Golden Rice part of the Generation Challenge programme) aims to improve nutrition and health by focusing on 'simple and efficient' screening of criteria for 'the micronutrient content of the seed' (see Brooks 2011, 185). The project involves iron-biofortified rice research, which is an example of the enhancement of staple crops through biological processes with the backing of the CGIAR as well as donors that include the Bill and Melinda Gates Foundation, the World Bank, the Syngenta Foundation, governments, research institutes, and several non-governmental organisations (see Ferroni and Castle 2011). What is noteworthy is that earlier attempts at biofortification had struggled for funding at the margins of the CGIAR. Until recently, trials of fortified varieties had been considered unsuccessful because of the complexity of the relationship of nutrition and health with the enrichment of staple crops. This changed when the testing of 'high-iron rice' in the Philippines was presented as having resulted in a 'biologically significant effect' in human subjects under experimental conditions. This was interpreted as a 'proof of concept' that requires further investments, similar to Golden Rice (see Brooks 2011, 185).

The problem with biofortification in the 1980s and 1990s was not solely that there was a shortage of investments; rather, the association of research into iron-rich rice with the MDGs was seen as side-lining the interactions among soils, crops, and human bodies. The likelihood of reaching similar experimental results elsewhere is therefore being overestimated when not including that the line of varieties needs to be cultivated under real-world conditions. Sally Brooks explains that the original testing of high-iron rice was accepted as successful because the experiment within which it was employed was 'designed to make it so' (ibid.). Subsequently, the experimental results became part of a policy agenda in which they were identified directly with the realisation of the MDGs and part of a programme that assumes a centralised, goal-driven research model wherein the materiality of rice is considered an obstacle to overcome (ibid., 186). In other words, a narrow focus on experimental results is turned into a project wherein crop scientists are working as part of global research communities and toward a new set of research goals that link nutrient targets to the MDGs. What this means is that the zinc content of a rice variety is privileged as a vehicle through which to improve health and nutrition at a global scale, which overestimates the ability to move from global successes to local circumstances.

This is also the case when considering the attempt to turn plant genetic diversity into a resource that can be used to create improved germplasm for breeding activities with selected crops (e.g., cassava, maize, rice, sorghum, wheat and tropical legumes, and cereals). This is the aim for another of the (Challenge) programmes of the CGIAR, which has partnered with scientists working on functional and comparative genomics in order to create databases that can be analysed with micro-arrays and molecular markers by multiple laboratories anywhere at any time. Hence, these crops can become a part of already ongoing comparisons of various crop genomes, and it would supposedly become possible to identify gene candidates and create a toolkit of genetic techniques that enable a more specific selection of genotypes.

Also in this instance, the experimental targets are selected in terms of their contribution to reaching a global target, and humanitarian licencing is used to ensure access to sophisticated techniques and the information that is produced. Accordingly, the CGIAR needs to enter into negotiations with companies and universities whose general policies are focused on patent applications. In this case, this means that the CGIAR must collaborate with partners who have their own patent strategies, which requires confidentiality to ensure that the novelty criterion for patents is not violated. This is significant since the involvement of partners has no limits at the level of the UN, while working with genetic diversity requires secrecy clauses to make it possible to work with the private sector, where patenting is the rule and the public domain is the exception.

Each of these examples shows how the ability to gain control over the agency of seeds is overestimated and the relationship to the environments in which seeds are planted and consumed is side-lined. Access is, then, not so much a solution to the problem of plant breeding and plant biotechnology requiring complex contractual arrangements with other stakeholders; the partnership agreements and humanitarian licences consolidate exceptions to patenting as strictly defined in terms of a group of beneficiaries of limited scope. It is not the complexity of who decides about the ownership and usage of seeds that is considered the aim in arranging access; rather, experts are mobilised to work on high-tech solutions to reach the MDGs as the centrepiece of joint rights and responsibilities in the name of the victims of malnutrition and those farmers who are expected to gain from cultivating this new category of high-performance crops.

## Conclusion

Our primary objective here was not a critique of the new 'miracle crops', the language of global targets, or even the new organisational model of global governance. Certainly these are features of the commodification of life, but what we wished to demonstrate is how species of plants are exemplary because of 'how' the transformation into species of life™ opens up a biopolitical horizon.

Increasingly, the exception and the incorporation of genetic resources into the framework of sovereignty, as its living body, are becoming indistinguishable. In this regard, Robinson Crusoe provides us with a suitable setting: an island that the sovereign individual inevitably leaves behind when a decision is forced upon him and his temporary reprieve between nature and society has ended. Such a metaphor of a no-man's land dramatises, in a sequential narrating of events, the moment of access and the loss of invention as an activity that is in harmony with nature. The story mirrors how creativity, entrepreneurship, and start-up culture are presented to us as a natural inclination of contemporary societies. One should keep in mind, however, that Crusoe could not remain in nature and rejoin the colonial world simultaneously: his fear for his life after seeing a footstep in the sand led him to stop innovating and turn to enclosure in defence. He saw this as the only course of action for a sovereign individual in nature, which was about to be overwhelmed by the uncivilised from the common world.

This is no arbitrary fictional counterpoint, juxtaposing a story to reality, intended to provide us with a convenient moral lesson. Rather, Defoe's story was much admired by Jean Jacques Rousseau, who modelled his state-of-nature theory on it. It was his favourite novel (Rousseau [1762] 1993b, 187), and it features in 'Emile' – a political treatise and an educational fantasy – as the only novel that his ideal pupil is allowed to read. The relationship is direct: Rousseau's child-savage had to be independent as long as possible, but 'to try to remain in it when it is no longer practicable, would really be to leave it, for self-preservation is nature's first law' (ibid., 187).[8] Obviously, this applies to Crusoe, giving up his inventions and leaving the island without having a choice in the matter. For Rousseau, however, it is as a fiction that his state of nature theory has the same status as the hypothesis of scientists.

As he explains it,

> the investigations we may enter into, in treating this subject, must not be considered as historical truths, but only as mere conditional and hypothetic reasoning, rather calculated to explain the nature of things, than to ascertain their actual origin, just like the hypothesis which our physicists daily form respecting the formation of the world.
>
> (Rousseau [1755] 1993a, 50, 51)

Natural philosophers of an earlier generation, such as Hobbes and Locke, had either entered into heated disputes with the first experimentalists or acknowledged the notion that science had a privileged relationship with nature. Taking a position in this regard reflected directly on the content of the social contract in its relations to the sovereign and his power over life and death (Shapin and Schaffer 1985; see chapter 2 on the significance of state-of-nature theories). Rousseau's solution is both elegant and provocative, claiming a similar privilege for his fictional state of nature, in the shape of Crusoe's adventures or those of his fictional pupil. Each of these enjoys the same relation to nature as the scientific hypothesis, which is to say that the birth of the modern novel (another of his pastimes) begins a long tradition of such fictional counterpoints. Soon after, Crusoe and Rousseau's savages turned into science-made monsters, chased by Doctor Frankenstein to the Arctic (Shelley [1818] 2006, 61). More recently there is Aldous Huxley's *Brave New World*, whose main character is named John Savage and who comes from the Savage Reservation. Eventually he chooses death and the horror of the last remnants of life in nature over the man–biology rationality that rules the world (Huxley 1967; Deibel 2009).

Our conclusion is not that there is a straight line from Crusoe via Rousseau to the Brave New World. Nor do we suggest that the contemporary life sciences are simply leading us into an age of instrumental reason or similar types of dystopian futures (Horkheimer and Adorno 1983). Rather, these fictions provide us with a counterpoint to how order is suspended through a 'fictitious lacuna'. Crusoe might be a character in a novel, but our reading helps us understand that the language of access is not simply about a number of exceptions. It has to do with a lacuna that is

> not within the law [*la legge*], but concerns its relation to reality, the very possibility of its application. It is as if the juridical order [*il diritto*] contained an essential fracture between the position of the norm and its application, which, in extreme situations, can be filled only by means of the state of exception, that is, by creating a zone in which application is suspended, but the law [*la legge*], as such, remains in force.
>
> (Agamben 2005, 31)

What Agamben is describing here is the contemporary predicament as situated in between life and law: a zone of indecision between, on the one hand, sovereignties that are neatly aligned through treaties that affirm a variety of claims and exceptions and, on the other hand, global concerns over plant materials given meaning and substance through their association with technoscientific win-win narratives. In turn, the essential fracture in the application of the norm is where this zone is situated, suspending its relation to reality, which we examined as a detachment of global environmental governance from environmental crisis in regard of the displacement and devaluation of the breeding of crops as a slow but steady accumulation of varieties and traits (Gepts 2004; Brush 2005a, 2005b).

This concept of the fictitious lacuna is part of his explanation of the global 'state of exception' (Agamben 2005). Accordingly, the state of exception is not strictly speaking about martial law, constitutional emergency powers, or a state of siege. Rather, Agamben argues that the state of exception is the dominant paradigm of government in contemporary politics (ibid., 2). He turns to Walter Benjamin to argue that this implies that there is an outside to law where the sovereign sanctions violence, and this implies an account of sovereignty where its power derives from the 'ultimate undecidability of all legal problems' (ibid., 55). Such indecisiveness is to be contrasted against the conventional view of the state of exception, which is preoccupied with how it is through making decisions that the sovereign maximises its own ability to exercise power.

The difference is fairly straightforward when we remind ourselves of how Robinson Crusoe is imagined as a sovereign individual in nature, where he is surrounded by savages coming to his personal island-kingdom. This formula is identical to the Hobbesian view of sovereign power as situated in a common world that is riddled with the fear of insecurity, whether through war, terror, failed states, or natural disasters. Contemporary equivalents are easy to find, including savage dictators with biological weapons, the threat of new genetic techniques falling into the wrong hands, or simply the intimate relationship between the life sciences and defence departments. Similarly, governments are constantly reminded that decisive action is needed in response to a wide range of insecurities, including environmental crisis, food shortages, and otherwise.

Our analysis, in contrast, foregrounds sovereigns becoming 'constitutively incapable of deciding' in the sense that such actions result in the suspension of the constitutive order that is the basis of the capacity to act (ibid., 31). The relationship of the sovereign power to nature is one of transgression, in the sense that its authority is overwhelmed in its application to reality – indecisive in its relation to the 'fundamentally unstable' relationship of the biological with the social (Pálsson et al. 2011). In our version, we meet the romantically defined scientists-altruists as they find themselves in a situation similar to that of early-modern colonial voyagers such as Crusoe. Also we only perceive glimpses of how a lasting legacy of deeply entrenched patterns of domination, dispossession, and exploitation is being created, inevitably as deeply invested in the tradition of Western biocolonialism and in relations of modern science and society (Kloppenburg 2005; Parry 2004; Bonneuil 2002).

To be precise, our analogy to Crusoe begins with a footstep set on a farmer's fields and a little later the enclosures (patents, UPOV 91, etc.) are in place. Increasingly there is a demand for protection from such insecurities, even though there were few buyers for genes with price tags attached. Plant genetic resources have been legally turned into materially bounded and discrete entities that can be stored, conserved, saved, secured, and so forth. Such a historical analogy affirms neither the responsible commoner nor the destruction of the commons. Rather, these perspectives co-exist within the context of the triangle of access, insecurity, and enclosure that is materialised in the shape of heavily negotiated exceptions for certain types of strategic resources mobilised around new commodity chains.

Accordingly, we see a global politics of access that enforces how staying out of the networks surrounding moral economies means 'not to win' and being excluded from the wider configuration of state/non-state interests and environmental/humanitarian causes. In this context, the usage of licencing and the appeal to social contracting is being imagined in terms of global targets and co-operative models that work in proximity to the experimental practises of the life sciences where DNA is expressed in digital or electronic form. For example, the CGIAR's activities have already showed how access is established as a feature of the usage of new genetic techniques and the need for cheap, efficient, and high-quality access to interactive software and databases on genetic diversity in informatic formats that can be downloaded, accessed, copied, searched, and/or circulated by researchers around the world. This is only one example, which opens up the biopolitical horizon of the increasingly complex and fragmented understanding of DNA-based code that should be understood in terms of the 'technological and economic management of information – that is, as a political economy' (Thacker 2005, 94; Pottage 2006).

The conclusion is therefore not that there is an 'island empire' being lost in the mist. Countless avenues (re-)appear for epistemological viewpoints that no longer rely on access, enclosure, and commons as pre-given categories, units of analysis, and fixed referents. This chapter has demonstrated exactly this in respect of the ambiguous moral work of high-tech forms of gifting, charity, and altruism – only through claiming a fictional counterpoint are we able to point out how heavily the advocates of access and closely related experts are invested in a future in which the commons has been detached from the conservation of threatened species of plant life on the basis of affirming cultural diversity as a basic condition for the protection of nature.

There is, of course, a next step. What remains, if one chooses to be really practical and realistic rather than engaging in plant branding, is to do the critical and interpretive work. Either we continue (re-)contextualising and (re-)materialising the common world in terms of how new and old biological entities can be mobilised in support of the lives of others or we leave any possibility of a genuine alternative behind, stuck in the laws of nature, frozen in time, and passive recipients of compensation schemes, bio-prospecting contracts, public–private partnerships, humanitarian licencing, and ever more grandiose social contracts.

## Notes

1 Of the big five seed/chemical companies, the only exception is BASF, which simply points to farming as the biggest job on Earth and to the importance of high yields (see

www.youtube.com/watch?v=6-oGEY4WJuI). The rest are typically sweeping in their ads or claims.

(1) Bayer 'leave a better world' (see www.youtube.com/watch?v=NtxwV4fzJSs).
(2) DuPont state that they realise potential (see www.youtube.com/watch?v=ufrSBEx QNmU).
(3) Dow's 2015 sustainability goals echo these ideas (www.dow.com/en-us/science-and-sustainability/2025-sustainability-goals/2015- sustainability-goals).
(4) Monsanto state that 'food is love' (see www.youtube.com/watch?v=U-jUytA-7ac).
(5) A Syngenta brand video uses the phrase 'bring plant potential to life' (see www. youtube.com/watch?v=fEgf6Y-Pry0). All links were current as of April 2016.

2 International undertaking on plant genetic resources for food and agriculture, 23 Nov. 1983, Art. 11. Resolution 8/83, from the 22nd Session, on 5–23 Nov. 1983. Material accessed on 1.12.2016 from www.fao.org/ag/cgrfa/IU.htm.
3 The examples are, in the order of the discussion in the paragraph:

(1) UNOP (2010) A Selection of Partnership Initiative (see www.un.org/partnerships/ partnership_initiatives.html accessed on 25.11.2010).
(2) UN CSD Partnerships Database (2004) Safe Water System Program (available at http://webapps01.un.org/dsd/partnerships/public/partnerships/94.html, accessed on 25.11.2010).
(3) UN CSD Partnerships Database (2008) Invasive Species Compendium Consortium, Available at http://webapps01.un.org/dsd/partnerships/public/partnerships/2354. html, accessed on 25.11.2010.
(4) Clear Air Asia annual report (accessed at http://cleanairasia.org/wp-content/uploads/ portal/files/documents/Clean_Air_Asia_Annual_Report_-_FINAL.pdf in December 2016).

4 See www.croptrust.org/ (accessed on May 2016).
5 Some articles are the following: Mellgren, Doug (27 February 2008) 'Doomsday' Seed Vault Opens in Arctic. From the Associated Press via msnbc.com. Retrieved on 3 July 2011, 'Noah's Ark' of Seeds Opens in Norway, by Elizabeth Weise, in *USA Today*. (See www.usatoday.com/tech/science/environment/2008-02-26-seeds_N.htm m). See also GRAIN's piece titled 'Faults in the Vault: Not Everyone is Celebrating Svalbard' (www.grain.org/es/article/entries/181- faults-in-the-vault-not-everyone-is-celebrating-svalbard, accessed on May 2016).
6 Ibid.
7 See www.planttreaty.org/'s Section 3, on innovative partnerships (accessed on 1.5.2016).
8 Rousseau's claim that men are naturally good is often misunderstood. Crucially, it applies only to man in nature until the encounter with others. Hence, 'to say that man by nature is good is to say that man was good while he remained natural'. Rousseau's man cannot be natural when living with others; therefore, 'he is corrupt and corrupts all he touches' (Cooper 1999, x, 2; see also Rousseau 1993b, 205).

## References

Agamben, G. (2005) *State of Exception*. Chicago: University of Chicago Press.
Allarakhia, M. and Wensley, A. (2007) Systems Biology: A Disruptive Biopharmaceutical Research Paradigm. *Technological Forecasting and Social Change*, 74(9): 1643–1660.
Andersen, N.A. (2008) *Partnerships: Machines of Possibilities*. Cambridge: Polity Press.
Bäckstrand, K., Khan, A., et al. (2010) *Environmental Politics and Deliberative Democracy*. Cheltenham: Edward Elgar.
Barry, K. (ed.) (2012) *Realism and Idealism in English Literature (Daniel Defoe – William Blake)*. Oxford: Oxford University Press.

Biermann, F., Pattberg, P., et al. (2007) Multi-Stakeholder Partnerships for Sustainable Development: Does the Promise Hold? In: Glasbergen, P., Biermann, F., and Mol, A. (eds.). *Partnerships, Governance and Sustainable Development: Reflections on Theory and Practice*. Cheltenham: Edward Elgar.

Boettiger, S. and Burk, D.L. (2004) Open Source Patenting. *Journal of International Biotechnology Law*, 1(1): 221–231.

Boisvert, V. and Vivien, F.D. (2012) Towards a Political Economy Approach to the Convention on Biological Diversity. *Cambridge Journal of Economics*, 36(5): 1163–1179.

Boyle, J. (2003a) The Second Enclosure Movement. *Law and Contemporary Problems*, 66(1): 33–74.

Boyle, J. (2003b) Enclosing the Genome: What Squabbles Over Genetic Patents Could Teach Us. *Advanced Genetics*, 50: 97–122.

Boyle, J. (2008) *The Public Domain: Enclosing the Commons of the Mind*. New Haven, CT: Yale University Press.

Brooks, S. (2011) Living With Materiality or Confronting Asian Diversity? The Case of Iron-Biofortified Rice Research in the Philippines. *East Asian Science, Technology and Society*, 5: 173–188.

Brush, S.B. (2005a) Protecting Traditional Agriculture Knowledge. *Washington University Journal of Law & Policy*, 17(1): 59–109.

Brush, S.B. (2005b) *Farmers' Rights and Protection of Traditional Agricultural Knowledge*. Capri Working Paper No. 36, International Food Policy Research Institute. Available at: www.capri.cgiar.org/ (accessed on 1.12.2016).

Burk, D. (2002) Open Source Genomics. *Journal of Science and Technology Law*, 8(1): 254.

Büscher, B. (2012) Payments for Ecosystem Services as Neoliberal Conservation: (Reinterpreting) Evidence From the Maloti – Drakensberg, South Africa. *Conservation and Society*, 10(1): 29–41.

Calvert, J. and Joly, P. (2011) How Did the Gene Become a Chemical Compound? The Ontology of the Gene and the Patenting of DNA. *Social Science Information*, 50(2): 1–21.

Carolan, M.S. (2010) Mutability of Biotechnology Patents: From Unwieldy Products of Nature to Independent Object/s. *Theory Culture & Society*, 27(1): 110–129.

Chander, A. and Sunder, M. (2004) The Romance of the Public Domain. *California Law Review*, 92: 1331–1371.

Cooper, M. (2008) *Life as Surplus: Biotechnology and Capitalism in the Neoliberal Era*. Seattle, WA: University of Washington Press.

Coupe, S. and Lewins, R. (2007) *Negotiating the Seed Treaty*. Warwickshire: Practical Action Publishing.

Defoe, D. (2007) *Robinson Crusoe*. Oxford: Oxford University Press.

Deibel, E. (2009) *Common Genomes: On Open Source in Biology and Critical Theory Beyond the Patent*. PhD dissertation. Available at: http://dare.ubvu.vu.nl/handle/1871/15441 (accessed on 1.12.2016).

Deibel, E. (2013) Open Variety Rights. *Journal of Agrarian Change*, 13(2): 282–309.

Deibel, E. and Mert, A. (2014) Partnerships and Miracle Crops: Open Access and Commodification in Agriculture and Food Production. *Asian Biotechnology Development Review*, 16: 1–33.

Drahos, R. (1996) *A Philosophy of Intellectual Property*. Aldershot: Ashgate Publishing.

Ferroni, M. and Castle, P. (2011) Public – Private Partnerships and Sustainable Agricultural Development. *Sustainability*, 3: 1064–1073.

Firn, R.D. (2003) Bioprospecting: Why Is It So Unrewarding? *Biodiversity and Conservation*, 12(2): 207–216.

Gepts, P. (2004) Who Owns Biodiversity, and How Should the Owners Be Compensated?' *Plant Physiology*, 134(4): 1295–1307.

Gepts, P. (2006) Plant Genetic Resources Conservation and Utilization: The Accomplishments and Future of a Societal Insurance Policy. *Crop Science*, 46(5): 2278–2292.

Gieryn, T.F. (1983) Boundary-Work and the Demarcation of Science From Non-Science: Strains and Interests in Professional Ideologies of Scientists. *American Sociological Review*, 48(6): 781–795.

Halffman, W. (2003) *Boundaries of Regulatory Science: Eco/toxicology and the Regulation of Aquatic Hazards of Chemicals in the US, England, and the Netherlands, 1970–1995*. Dissertation. Amsterdam: University of Amsterdam.

Hamilton, C.J. (2006) Biodiversity, Biopiracy and Benefits: What Allegations of Biopiracy Tell Us About Intellectual Property. *Developing World Bioethics*, 6(3): 158–173.

Haraway, D.J. (1997) *Modest_Witness@Second_Millennium.FemaleMan©_Meets_OncomouseTM: Feminism and Technoscience*. New York, NY: Routledge.

Hardin, G. (1968) Tragedy of the Commons. *Science*, 162(3859): 1243–1248.

Harvey, D. (2011) The Future of the Commons. *Radical History Review*, 109: 101–107.

Hatanaka, M. (2009) Certification, Partnership, and Morality in an Organic Shrimp Network: Rethinking Transnational Alternative Agrifood Networks. *World Development*, 38(5): 706–716.

Hayden, C. (2003a) From Markets to Market: Bioprospecting's Idiom of Inclusion. *American Ethnologist*, 30(3): 1–13.

Hayden, C. (2003b) *When Nature Goes Public: The Making and Unmaking of Bio-Prospecting in Mexico*. Princeton, NJ: Princeton University Press.

Heller, M. (1998) The Tragedy of the Anticommons: Property in the Transition From Marx to Markets. *Harvard Law Review*, 111(3): 621–688.

Heller, M. and Eisenberg, R. (1998) Can Patents Deter Innovation? The Anticommons in Biomedical Research. *Science*, 280(5364): 698–701.

Helmreich, S. (2008) Species of Biocapital. *Science as Culture*, 17(4): 463–478.

Helmreich, S. (2009) *Alien Ocean: Anthropological Voyages in Microbial Seas*. Berkeley, CA: University of California Press.

Hess, C. and Ostrom, E. (2003) Ideas, Artifacts, and Facilities: Information as a Common Pool Resource. *Law and Contemporary Problems*, 66(1–2): 111–146.

Hisano, S. (2005) A Critical Observation on the Mainstream Discourse of Biotechnology for the Poor. *Tailoring Biotechnologies*, 1: 81–106.

Hoare, A.L. and Tarasofsky, R.G. (2007) Asking and Telling: Can 'Disclosure of Origin' Requirements in Patent Applications Make a Difference. *The Journal of World Intellectual Property*, 10(2): 149–169.

Hope, J. (2008) *Biobazaar: The Open Source Revolution and Biotechnology*. Cambridge, MA: Harvard University Press.

Hope, J. (2009) Open Source Genetics: Conceptual Framework. In: Overwalle, G. (ed.). *Patent Pools, Clearinghouses, Open Source Models and Liability Regimes*. Cambridge: Cambridge University Press.

Horkheimer, M. and Adorno, T.W. (1983) *Dialectic of Enlightenment*. London: Allen Lane.

Hughes, S. and Deibel, E. (2006) Plant Breeder's Rights: Room to Manoeuvre? *Tailoring Biotechnologies*, 2(3): 77–86.

Huxley, A. (1967) *Brave New World*. London: Penguin Books.

Kaan, C. and Liese, A. (2010) Public Private Partnerships in Global Food Governance: Business Engagement and Legitimacy in the Global Fight Against Hunger and Malnutrition. *Agriculture and Human Values*, 28: 385–399.

Kloppenburg, J. (2005) *First the Seed: The Political Economy of Plant Biotechnology*. Madison, WI: University of Wisconsin Press.

Kryder, R., David, R., et al. (2000) The Intellectual and Technical Property Components of Pro-Vitamin A Rice (GoldenRice™): A Preliminary Freedom-to-Operate Review. *ISAAA Briefs*, No. 20. Ithaca, NY: ISAAA.

Lee, D. and Wilkinson, R. (eds.) (2007) *The WTO After Hong Kong: Progress in, and Prospects for, the Doha Development Agenda*. New York, NY: Routledge.

Lesser, W.H. and Krattiger, A. (2007) Valuation of Bioprospecting Samples: Approaches, Calculations, and Implications for Policy-Makers. In: Krattiger, A., et al. (eds.). *Intellectual Property Management in Health and Agricultural Innovation: A Handbook of Best Practices*. Davis, CA: PIPRA.

Marx, K. (1990) *Capital: A Critique of Political Economy, Volume 1*. London: Penguin Group.

May, C. (2000) *The Global Political Economy of Intellectual Property Rights: The New Enclosures?* London: Routledge.

May, C. and Sell, S.K. (2006) *Intellectual Property Rights: A Critical History*. Boulder, CO: Lynne Rienner Publishers.

Mert, A. (2009) Partnerships for Sustainable Development as Discursive Practice: Shifts in Discourses of Environment and Democracy. *Forest Policy and Economics*, 11(5–6): 326–339.

Mert, A. and Chan, S. (2012) The Politics of Partnerships for Sustainability Development. In: Pattberg, P., et al. (eds.). *Public – Private Partnerships for Sustainable Development: Emergence, Influence and Legitimacy*. Cheltenham: Edward Elgar.

Mooney, P. (1979) *Seeds of the Earth: A Private or Public Resource?* Ottawa: Inter Pares.

O'Malley, M. and Dupré, J. (2007) Size Doesn't Matter: Towards a More Inclusive Philosophy of Biology. *Biology and Philosophy*, 22(2): 155–191.

Opderbeck, D.W. (2004) The Penguin's Genome, or Coase and Open Source Biotechnology. *Harvard Journal of Law & Technology*, 18(1): 167–227.

Ostrom, E. (1990) *Governing the Commons: The Evolution of Institutions for Collective Action*. Cambridge: Cambridge University Press.

Overwalle, G. (ed.) (2009) *Patent Pools, Clearinghouses, Open Source Models and Liability Regimes*. Cambridge: Cambridge University Press.

Pálsson, G. and Prainsack, B. (2011) Genomic Stuff: Governing the (Im)Matter of Life. *International Journal of the Commons*, 5(2): 259–283.

Parry, B. (2004) *Trading the Genome: Investigating the Commodification of Bio-information*. New York, NY: Colombia University Press.

Pattberg, F., Biermann, S., et al. (eds.) (2012) *Public – Private Partnerships for Sustainable Development: Emergence, Influence and Legitimacy*. Cheltenham: Edward Elgar.

Pottage, A. (2006) Too Much Ownership: Bioprospecting in the Age of Synthetic Biology. *BioSocieties*, 1(1): 137–158.

Poulton, C. and Macartney, J. (2012) Can Public – Private Partnerships Leverage Private Investment in Agricultural Value Chains in Africa? A Preliminary Review. *World Development*, 40: 96–109.

Pufendorf, S. (1991) *On the Duty of Man and Citizen According to Natural Law*. Cambridge: Cambridge University Press.

Reid, W.V., Laird, S.A., et al. (1993) *Biodiversity Prospecting*. Washington, DC: World Resources Institute.

Rousseau, J.J. (1993a) *The Social Contract and Discourses*. London: Everyman.

Rousseau, J.J. (1993b) *Émile*. London: Everyman.

Safrin, S. (2004) Hyperownership in a Time of Biotechnological Promise: The International Conflict to Control the Building Blocks of Life. *The American Journal of International Law*, 98(4): 641–685.

Sell, S. (2007) Intellectual Property and the Doha Development Round. In: Lee, D. and Wilkinson, R. (eds.). *The WTO After Hong Kong*. London: Routledge.

Shackelford, S.J. (2009) The Tragedy of the Common Heritage of Mankind. *Stanford Environmental Law Journal*, 109: 102–155.

Shapin, S. and Schaffer, S. (1985) *Leviathan and the Air-Pump: Hobbes, Boyle, and the Experimental Life*. Princeton, NJ: Princeton University Press.

Shelley, M. (2006) *Frankenstein; or, the Modern Prometheus*. London: Penguin Books.

Spielman, D.J., Hartwich, F., and von Grebmeri, K. (2010) Public – Private Partnerships and Developing Country Agriculture: Evidence From the International Agricultural Research System. *Public Administration and Development*, 30: 261–276.

Sunder, M. (2007) The Invention of Traditional Knowledge. *Law and Contemporary Problems*, 70: 97–124.

Tamminen, S. and Brown, N. (2011) Nativitas: Capitalising Genetic Nationhood. *New Genetics and Society*, 30: 73–99.

Thacker, E. (2005) *The Global Genome: Biotechnology, Politics and Culture*. Cambridge, MA: MIT Press.

van Berloo, R., Van Heusden, S., et al. (2008) Genetic Research in a Public – Private Research Consortium: Prospects for Indirect Use of Elite Breeding Germplasm in Academic Research. *Euphytica*, 161(1–2): 293–300.

van Dooren, T. (2007) Terminated Seed: Death, Proprietary Kinship and the Production of (Bio)Wealth. *Science as Culture*, 16(1): 71–94.

van Dooren, T. (2009) Banking Seed: Use and Information in the Conservation of Agricultural Diversity. *Science as Culture*, 18(4): 373–395.

Wield, D., Chataway, J., and Bolo, M. (2010) Issues in the Political Economy of Agricultural Biotechnology. *Journal of Agrarian Change*, 10(3): 342–366.

Yoxen, E. 1984. *The Gene Business: Who Should Control Biotechnology*. New York, NY: Harper & Row.

# 5 Animal genetic resources as a global matter of concern[1]

Over the last decade, animal genetic resources (AnGRs) have become a topic of renewed interest in the international politics that takes agricultural species as its object of concern. We can point to several reasons for this, which are mutually intertwined. The first has to do with the unclear legal status and the scope of regulation stemming from biodiversity agreements that target a wide range of animal species from wild to agricultural, their biological materials, and various genetic resources within the global politics of the late twentieth and early twenty-first centuries.

Here, states, nations, and indigenous communities have newly become key stakeholders of genetic resources, as they have been granted sovereign rights over territorially bound, native '*in situ*' resources via the text of the Convention on Biological Diversity,[2] signed by more than 150 states in 1992. The sovereign rights over genetic resources (GRs) have since been re-enforced with the adoption of the Nagoya Protocol on Access and Benefit-sharing entering into force on October 2014. The Nagoya Protocol is a legally binding protocol guiding in how to interpret and act on the genetic-resources issues presented in the CBD more than twenty years earlier. The second of the inherently nested interests is related to the way in which access and animals could or should be regulated internationally within biodiversity frameworks. Debate can centre on how various kinds of genetic resources, from plants to animals and from wild to agricultural species, differ from each other and how the differences may have to inform the practical execution of their global governance.

The Food and Agriculture Organization of the United Nations established the Commission on Plant Genetic Resources in 1983 to deal with policy, access, and benefit-sharing issues related to plant genetic resources. The FAO did broaden the mandate of the Commission in 1995 to cover all aspects of it after the CBD entered into force and after recognising 'that broadening the coverage of the Commission would allow the Organization to deal in a more integrated manner with agrobiodiversity issues' (FAO 1995, 66). Two years later, in 1997, the Commission also established separate working groups for animal and plant genetic resources, followed by one expert group for forest genetic resources. All these committees – specifically established for different types of genetic resources – demonstrate how difficult the policy, ownership, and access- and benefit-sharing issues related to GRs, especially AnGRs, are to understand, let alone to manage in practise.

Three examples from international analyses from the last ten years will clarify some of the tricky aspects of global agreements and governance of agricultural AnGRs and will open up good questions as to how AnGRs became such politically contested objects of agricultural nature.

First, we can consider a report from 2006 exploring policy options for the 'Exchange, Use and Conservation of Animal Genetic Resources', commissioned by the FAO and funded by the Government of the United Kingdom of Great Britain and Northern Ireland. This recognised that a fundamental tension between the traditional ownership of AnGRs and new global conventions had emerged (Hiemstra et al. 2006), and this tension needs to be resolved on the international level. Going through various options for AnGR regulation, tellingly the report ended with a summarising paragraph claiming that '[c]lassical ownership' of AnGR includes physical ownership and communal 'law of the land' affecting livestock keeping and breeding. There is an increasing tension with developments in the realms of biodiversity law and intellectual property rights protection. Demarcation of these different rights systems and maintaining equity among different stakeholders is crucial to avoiding conflict and increased transaction costs. In this context, it is important to consider the rights of livestock keepers/breeders vis-à-vis national-level sovereign rights, as well as obligations between patent holders and breeders/livestock keepers (Hiemstra et al. 2006, 37).

The report had been commissioned because the FAO wanted to clarify the options for navigating the post-CBD world of new political and legal frameworks for AnGR management. Three years later, another expert report on AnGRs raised a concern that is related to the fact that all types of ownership relations now faced a potential disruptive element:

> Private or communal ownership of AnGR, is potentially at least, challenged by national sovereignty over genetic resources. Individual owners may find that their rights to sell breeding animals or other genetic material, particularly across national boundaries, are restricted. Those seeking to buy specific AnGR may find that they are unable to do so, or that they can only do so on terms that are acceptable not only to the owner of the resources but also in compliance with national legislation.
>
> (Commission on Genetic Resources for Food and Agriculture 2009, 29)

Finally, in November 2014, the Intergovernmental Technical Working Group on Animal Genetic Resources for Food and Agriculture concluded after its meeting that more work on AnGRs is needed but that at least the types of their **utilisation**, the criteria and approaches in assessment of the **country of origin** of AnGRs and **access- and benefit-sharing policies** all need further clarification, although at global level several internationally binding legal treaties do exist (Commission on Genetic Resources for Food and Agriculture 2014, 19–25).

For a long time, animals, breeds, and their genetic resources were governed solely via rights that were based on physical access and use rights to animals, as animals and breeds were seen as 'wholes', whether living animals or recorded breeds. A mix of private, semi-private, and common ownership models for agricultural and farm animals has been in use, and these have also generated much discussion about the forms of entitlement over the life of the animals and the best possible ways to organise these relations (Hardin 1968; David 2011). However, as the biotechnologies used in animal production have developed – increasing animal growth rates and affecting carcass composition, enhancing disease-resistance, and improving hair and fibre production (Wilmut et al. 1992; Wheeler et al. 2010; Wheeler 2013) – the value of an individual farm animal, or even that of a breed, has become calculated not just in direct relation to the output of agricultural goods (e.g., meat or milk) but also by its value within the social system of breeding. Thus, farm animals and breeds have become valuable also for their capacity to produce particular kinds of offspring or to transmit valuable features encoded within their DNA. This capacity can be codified either in rough ideas of maintaining a pure breed type or within sophisticated algorithms used on modern farms for calculating the 'estimated breeding value', which are based on the development of the herdbook, an innovation that enables population management through precise recordings made in the late eighteenth and early nineteenth centuries.

Given that the two sources of value in farm animals have been recognised for more than 200 years, it is surprising that at present, the global community dealing with AnGRs has ended up in a situation wherein matters of access, ownership rights, and benefit-sharing – issues that for a long time remained unchallenged – have become a source of global concern and of slowly proceeding political processes for which there are no easy resolutions. The 'sovereignty' over all types of genetic resources as described in the major international agreements gives the signatory states relatively free hands to develop and implement national laws and regulations. In fact, to fulfil their sovereignty over AnGRs, for example, the states must decide what types of entitlements and relationships to AnGRs they should implement and how this is related to the national rights that farmers have over their animals, for example.

This chapter continues the analysis in the last chapter by examining two questions about the nature and status of the AnGRs in the post-CBD world. First, why are the issues of AnGRs for agriculture debated alongside more general issues of biological diversity? Are the animals not reducible to biological diversity? Second, how should we understand the idea of sovereignty over genetic resources given to the signatory states of the CBD and the Nagoya Protocol in the context of AnGRs? The answers to these questions are painting a picture of novel global biopolitical terrain, at once dominated by the idea of 'sovereign' and by ideas of nature that point to it as being mankind's common heritage for open access by all.

The first question builds on an institutional trajectory – an institutional metacode – manifested through the cultural history of AnGRs' movement in political institutions, most notably the FAO. We claim that early warnings about the need for conservation and coordinated management of AnGRs for agriculture did not lead to action; instead, they failed to mobilise larger communities to action. This, in turn, led the animal geneticists affiliated with the FAO and other interested parties to join forces with the environmental conservation movement, especially the United Nations Environmental Programme (UNEP), for soliciting international support for responding to the issue of conservation, then seen as an agenda priority.

Second, and following the first institutional trajectory, the way in which the CBD and to certain extent the subsequent Nagoya Protocol defined and understood genetic resources owes much to the world of plant genetic resources (PGRs). Defining the rights and obligations of signatories through PGRs leads implicitly to the world of plant breeding, which operates differently from that of animal-breeding practises, the key economic relations, and related biological processes. This is also why the key articles and provisions of the biodiversity conventions are couched in strong terms under national governments' sovereign powers. This is the second metacode element for understanding why animals are turned into genetic resources and why ownership issues are so hard to resolve in relation to questions about the animal.

This has resulted in a world in which we have moved from a system in which animals were part of a seamless universal nature, without political boundaries, to a world that is a collection of discrete 'national genetic landscapes' safeguarded by state policies and legal provisions and by the 'sovereign'. This is the new biopolitical terrain explored in the previous chapter. The difference, however, is that this chapter focuses on the transformation of farm animals and breeds into nationally recognised genetic resources, approached again via an institutional exploration of incommensurable biopolitical philosophies. The latter have resulted in a global but fragmented epistemic space implied or even formed by the metacodes, comparable yet different, as will be explained in detail.

## The question of the animal and its institutional trajectory

The management of farm AnGRs became topical immediately after the establishment of the United Nation's Food and Agriculture Organization; however, the concerns related to genetic resources were identified initially as a challenge for developing countries. The negative consequences of modern animal production aimed at increasing animal productivity started to raise doubts within the scientific community, and the first calls for genetic conservation followed quickly. Ralph W. Phillips, the first Deputy Director-General of the FAO, remembers how he, as the first employee of the animal section of the FAO, had for his whole career 'already carried out activities relating to animal genetic resources [. . .] and the Organization's involvement in this work dates back to 1946' (Phillips 1981, 5). From early on, the worry was about losing local breeds to extinction in developing countries. Local animals were replaced with globally homogenised, more productive breeds that became easily available and were adopted at a fast pace. Despite the early warning calls, little to no action aimed at conservation ensued at global level even if the FAO did produce a number of scientific reports and hosted a series of meetings about the issue between the early 1950s and the 1960s.

It was only after the widespread negative impacts of the Green Revolution became evident in the 1960s that AnGRs became truly a global matter of concern also for scientists working in developed countries. This was the direct result of unmanaged use of new breeding techniques in combination with shrinking and homogenised ecological habitats. For example, at the 1969 regional meeting of the European Association for Animal Production, the issue of 'gene pool losses' had been clearly articulated by Kalle Maijala (1971), who also identified the root cause for these losses: 'The present era of frozen semen [. . .] has reactualized the problem of gene losses [. . .]. The problem arises mainly from the fact that an effective utilization of the best animals of today automatically means setting aside the poorer animals, strains, breeds and even species' (ibid., 403–444).

In response to these developments, the FAO and UNEP launched a joint project in 1974 with the title Conservation of Animal Genetic Resources. It had as its key objective to 'prepare a list of breeds of farm animals in danger of extinction together with an account of any measures which have been recommended or taken to prevent this extinction' (Mason 1981, 17). A Consultation Report followed (Mason 1981), with a review of the work achieved by the project through the participating regional and national organisations. This made recommendations for future action.

The report was presented at a workshop for animal geneticists working with genetic resources and was framed with opening words from Phillips that simultaneously exhibited hope and exasperation with the current state of affairs. He proclaimed:

> I am pleased to bid you welcome here, on behalf of the Director-General. It is indeed heartening to see such a distinguished group of animal geneticists assembled to consider the problems of identification, conservation, and effective management of animal genetic resources. This is a matter critical to man's future, yet it has had little recognition and little real attention.
>
> (Phillips 1981, 2)

This opening speech betrays how, by the early 1980s, while the animal scientists had been awakened to the dire straits of genetic resources, political support for addressing the issue was still weak. More generally, the matter was still unrecognised as a global political issue. It did not appear on general global agendas as did other issues related to modernisation and increased production, such as those targeted by the environmental movement, which had already in the 1970s started to attract more political attention and rapidly gain political weight in the international political arenas. As a result, the issue of farm AnGRs did not spur action or attract funding for conservation efforts (Boyazoglu and Chupin 1991).

Fast-forward a decade to the early 1990s, and one finds more explicit frustration with the slow progress of conservation efforts and the lack of coordinated international action. Explaining the issue and the need for AnGR conservation (Hodges 1990), a senior officer with the FAO wrote that 'the time for technical talk is over. The issues are clear. What is now needed is an effective international decision to provide funds to do what all agree is now necessary [at] the global, regional and national levels' (Hodges 1990, 153). AnGRs needed more political support, but this proved to be hard to gain without rethinking and reframing the issue and joining forces with other institutional actors. International action did finally follow a few years later, in 1992, when the FAO joined forces with the UNEP and co-organised the Rio Earth Summit in Rio de Janeiro, Brazil. This was a historical moment for AnGRs. The summit was the place and the time at which genetic resources became newly articulated as parts of nature, as they were linked directly to the recently introduced concept of biodiversity, the key theme of this global meeting for the world's leaders.

At the meeting, the UNEP and FAO introduced the global CBD, a convention aimed at saving biodiversity, to the larger public and opened it for signatures. It was signed by some 160 countries in Rio de Janeiro, and more than 30 more followed suit in the next few years. Several of the articles in the Convention addressed the issue of genetic resources directly and introduced an obligation to identify, report, and take appropriate actions to conserve genetic resources. The lengthy follow-up work finally resulted in the Global Plan of Action for Animal Genetic Resources (GPA), adopted in 2007, and the Guidelines on the Preparation of National Strategies and Action Plans for AnGRs, published in 2009.

Given the half-century history of AnGRs as a matter of concern for animal geneticists and the long wait for political action, the key question is why the broad political traction to save genetic resources emerged only with the introduction of the CBD, leading to the global and national action plans and guidelines specific to AnGRs over a decade later.

## Early failures in valuation

The reason the FAO and regional institutions such as the European Federation for Animal Science (EAAP) failed in gaining political traction with their early alerts about the need for conservation measures is related to two shortcomings in the definition and the valuation of AnGRs.

The first shortcoming was the lack of consensus in scientific definition, valuation, and prioritisation of AnGRs that could lead to simple and uniform action recommendations. The question of what it is exactly that needs to be conserved and how to prioritise the required conservation actions was left open, or at best was illustrated through cases of a few particular breeds. The more important shortcoming was the failure in global political and legal identification of the responsible parties and beneficiaries of any value derived from the costly conservation actions. This, in turn, is linked to the fact that, until the CBD, AnGRs had been treated as a mixture of 'private' and 'commons', or as 'club commons' (David 2011) to be shared and used, subject only to individual farmers' and breeding associations' property-right regimes and explicit regulations at country level.

After the introduction of the CBD, the legal status of AnGRs changed globally: they were politically identified as falling under the sovereign power of the signatory parties to the Convention. This marked a major change and complication in access- and benefit-sharing relations that was later affirmed by the GPA in 2007 and later by the Nagoya Protocol. Understanding the latter is especially important, since this understanding exposes the new overarching paradigm under which the value of most AnGRs today is to be governed.

The failure to provide a clear direction for conservation is related to the arguments about the overall role of various kinds of AnGRs in animal production. When the concept of AnGRs was first introduced among animal scientists, they were framed in and through two distinct means (both scientifically informed) for demonstrating the role of AnGRs in animal production. In both means – involving 'utilizationist' and 'conservationist' standpoints – the value of AnGRs in animal production was literally assigned in two incommensurable ways (and to some extent this debate continues even today). The earlier-mentioned report on genetic resources explained the main differences between the two approaches:

> The utilizationist's primary concern is the immediate usefulness of available genetic resources to improve livestock populations [. . .]. The loss of breeds as distinct identities is not generally a concern, as long as the genes that make these breeds potentially useful are retained in the commercial stocks . . . The preservationist's primary objective is long-term conservation of genetic resources for future use. This view emphasizes the value of preserving the widest possible spectrum of genetic diversity to be prepared for unpredictable changes of future needs. The greatest possible number of breeds are to be preserved as purebreds.
>
> (Hodges 1984, 2–9)

The differences in these two views boil down to conserving 'the known useful genes' in one form or another versus conserving the 'genetic diversity of whole animal breeds' to hedge the uncertainty stemming from unknown future needs. The first approach is intended to save the sliced and diced, functionally valuable component of animals, no matter the 'breed'; the other involves also the animal breeds in pure-bred form and maximising diversity as an insurance policy against future unknowns. Although it is analytically distinct from animals or breeds, the animal scientists first presented the issue of AnGR conservation as a choice between **isolated genetic components** of immediate utility in the production of high-performance animals and the maximisation of genetic diversity via conservation of **local breeds in their animal forms**. In these two approaches, AnGRs are conceptually presented as different objects of conservation and seen as valuable for different purposes.[3]

In addition, the failure to identify the parties responsible for conservation irked conservationists, since this is related directly to the economics of conservation or, more generally, to the political economy of global animal production. The problem was captured in a report produced by the United States Board of Agriculture National Academy of Science (1993, 3):

> The concept of conservation [. . .] is complex. One can think of live animals, being preserved in situ, or in some semi-artificial situation; alternatively one may think of cryogenic storage of sperm or fertilized ova or other tissues or gene segments. The economic problems are difficult with both live animals and with haploid or diploid cells. Who is to pay? There are also questions of how many to preserve, for how long, and where.

Although plant varieties and their genetic material had been protected by various intellectual-property systems since the 1930s (see Kloppenburg 1988), AnGRs were used by farmers and breeder associations alike without generalised or specified rights and restrictions imposed at global level. Since there was no definition of the ownership rights to genetic materials of animals, the global assignment of conservation responsibility through political processes proved to be impossible without more specific consideration. Yet the ownership and responsibility questions have been more straightforward with regard to pigs and chickens. This reflects what M. Tvedt et al. (2007, 8) note about the legal protection of farm animals (and their protection in general) and of chickens and pigs in particular:

> 'For farm animals there are strong biological and physical means of protection available: The owner of the animal can more easily than the plant breeder have an overview and control over who is receiving genetic material from his animals or his population. For poultry and pig breeding, however, where farmers often buy hybrids whose genetics are more difficult to reproduce. The sale of hybrids is thus an important strategy for maintaining physical control over the genetic material by physical control over the material. For other breeds, in particular cattle, the physical ownership is often combined with a register, a herd book that maintains a protocol for the generations of animals fulfilling the criteria for registration'.

This is why the various claims about the value of AnGRs and the need for their conservation, made by both the 'utilizationists' and 'conservationists', rang on deaf ears outside animal-scientist circles. The failure to spur action was not born out of the scientific challenge to demonstrate the value of AnGRs in animal production or the lack of consensus on setting the priorities for conservation. Instead, and above all, it was a problem of political economy: who is to pay? And even more importantly, who is to benefit?

### *Global reframing of animal genetic resources*

The Food and Agriculture Organization remained active on AnGRs since the FAO/UNEP consultation programme in 1980; established its Committee on Agriculture, which kept reminding about the issue at the FAO Council level; and designed an FAO expert consultation round on AnGRs in 1989 and in 1992 (FAO 1990, 1999; Steane 1992). What became clear over the years was that a global binding framework was needed.

Anticipating the global political agreement on AnGRs, Hartwig de Haen (1992), Assistant Director-General of the Agriculture Department of the FAO, wrote in 1992 that

> it is clear that there is a greater awareness that a framework for the management of global animal genetic resources must be established. It is most appropriate that this Expert Consultation is taking place now in the context and

> timing of the Earth Summit, the United Nations Conference on Environment and Development (UNCED) to be held in Brazil in about eight weeks' time.
> (de Haen 1992, 3)

The first reframing of the AnGRs came in the form of the global CBD a few months later. The Convention had been long in preparation, and the FAO had been involved in its drafting phases, influencing, among other issues, the inclusion of genetic resources in the text and their definition. There were two important reframings in the Convention. First was the definition of the genetic resources, as genetic material of 'actual or potential value' (CBD, Article 2). This definition bridged the two views on the valuable material to conserve, or the 'utilizationist' and 'preservationist' standpoints. Genetic resources become genetic material that can be ascribed demonstrable or imaginable value. But the question then arises of who has the right to link any value claims to AnGRs.

The other reframing answered this question. Under the definitions in Article 2 and Article 15, genetic resources found *in situ* within the territory of a signatory were identified as belonging under the sovereign power of signatory states representing the nations of the world, reframing their ownership relationships globally. This is how the CBD enacted an important political redefinition of genetic resources: previous problems with definition of the value of non-human life were re-articulated through the politics of nationhood, in the idea of national differences found within the CBD's vision of genetic nature.

With the CBD, also AnGRs became tightly nested within the sovereignty of nation-states and their geography. A reversal of the old idea of nations being rooted in natural differences of human populations took place – non-human populations, conceptualised as 'genetic resources', could now be identified and placed under national or international jurisdiction in terms of their geographical location and the political powers representing the nationhood that governed that geographical area. A global cartographic demarcation of non-human life took place as these novel objects of nature were grafted to the foundations of national sovereignty. They became a new part of the body of nations, a novel form of non-human nationhood.

The Convention assumes that a significant amount of power over AnGRs and their governance rests with the signatory nation-states. Tvedt et al. (2007, 24) interpret the Convention and its provisions in the following manner:

> The CBD presupposes the right of a country to exercise sovereign control over its AnGR (accompanied by a number of responsibilities). From the perspective of an exporting country, one of its main concerns is to maintain any property rights it may wish to retain over the AnGR after the resources have left the country. Similarly, it may wish to ensure that the rights of the exporter are respected by the buyer/importer of the AnGR. The most prominent rationale for a country to regulate export of AnGR would be to secure a right over that particular material in the future, including preventing that countries or companies gain control over these resources (e.g., through patenting or other

forms of intellectual property rights), which might reduce the value of it in the exporting country.

This reframing introduced a whole new system wherein the value of any animal breed will be decided by the signatory nations but without any common reference as to what constitutes a legitimate value claim over the material, except the condition '*in situ*'. In the CBD, this refers to the 'conditions where genetic resources exist within ecosystems and natural habitats, and, in the case of domesticated or cultivated species, in the surroundings where they have developed their distinctive properties' (CBD, Article 2). The *in situ* condition for valuable genetic resources has had tremendous effects on how AnGRs are seen in the post-CBD world, especially because AnGRs thereby became divorced from the idea of being freely circulated or tradable objects of nature. They stopped being global commons and instead became subject to the political powers of the Convention states, many of whom did not have and still do not have a clear stance on what are 'valuable' AnGRs to them and on how they will enact their sovereign powers over access and benefit sharing for the valuable AnGRs. Definitional and legal disorientation followed.

The third reframing of AnGRs happened as they were presented through ideas derived from the plant and crop worlds. The FAO background study for the CBD on the 'Exchange, Use and Conservation of Animal Genetic Resources' acknowledged this as a major problem. It explained that, although current debates regarding agricultural genetic resources have largely had a crop/plant focus, these discussions and the international instruments or agreements that are emerging have tended to frame the debate for AnGR as well. At first sight, plant breeding does not differ much from animal breeding. The genetics of plants and animals are based on the same principles. Plant and animal breeders both need genetic diversity in order to advance, and the genetics determine adaptation to particular agro-ecological circumstances, as well as product qualities to a large extent. However, plant varieties can be protected by a plant breeder's rights (UPOV), which is not the case for animal breeds/strains. Plant breeders aim at the development of new uniform varieties that are defined by certain phenotypic traits that can identify them from other varieties. Farm animal breeding is largely based on the selection of individuals within populations rather than selection between populations or strains. Farm animal breeders are interested in individual animals (within populations/breeds), while the whole population of a plant variety (clones) is the main focus of plant breeders (Hiemstra et al. 2006, 22).

The third reframing, then, pointed to the difference between animal and plant genetic resources as biological bred resource and as legal protected asset: animals might carry interesting genetic traits, but it is difficult to exploit one unique genetic characteristic. There are no large international breeding centres. Most breeding takes place on farms – except for poultry and, in some cases, pigs – and the centres of origin or diversity for AnGR are not as clearly defined as for plants. Most importantly, farmers are not protected by internationally binding rights frameworks, while plant breeders are, by the International Union for the

Protection of New Varieties of plants (UPOV).[4] The differences between PGRs and AnGRs make it hard to enforce a single system for the two; however, the CBD does exactly this by enforcing the sovereignty of the signatory states as its starting point for rights and obligations via discourses that are appropriate mostly for plant genetic resources.

These reframings of the AnGRs dictate much of how global action now unfolds. Fifteen years after the CBD, in 2007, the state representatives adopted the first GPA at the Interlaken Conference, held in Switzerland, in what was called a 'historical breakthrough' by FAO Director-General Jacques Diouf (FAO 2007, iii). This GPA includes the 'Interlaken Declaration on Animal Genetic Resources', in which the sovereign right of states over their AnGRs for food and agriculture was restated (in declaration point 2).

## In situ, *transboundary, and domestic applications*

The fact that animals can move across politically established boundaries created a potential problem with regard to these sovereign rights, however, and led to new politically innovated categories of AnGRs, such as 'transboundary' for species that criss-cross institutionalised national borders. The GPA explained:

> Assessing the status of animal genetic resources on a global scale presents some methodo-logical difficulties. In the past, analysis of the Global Databank to identify breeds that are globally at risk was hampered by the structure of the system, which is based on breed populations at the national level. To address this problem [. . .] a new breed classification system was developed. Breeds are now classified as either local or transboundary, and further as regional or international transboundary.
>
> (FAO 2007, 13)

With these political documents, not only did animals considered to be genetic resources become 'national' (pertaining to a state), but some of them also became 'transboundary', regionally and internationally. A result of this is that political categories are infused with conservation-science categories because of the political economy involved in the ownership rights over the actually or potentially valuable genetic resources.

These categories are as much politically informed as they are scientifically valid. The definitions of '*in situ*' or 'transboundary' are inherently related to the political cartographic demarcation of the natural ecologies of domesticated animals, pointing to the deep connection between politics of value and the science of conservation of farm animal genetic resources. This is what eventually created the incentive for nation-states to act on the issue of genetic erosion of animal populations, but it is now, at the same time, generating new challenges, which it is beyond the ability of animal scientists or even international organisations to resolve.

This complexity is reflected in how national legislation has been drafted and implemented. Writing about the challenges in the implementation of the CBD, legal experts Matthias Buck and Clare Hamilton claim that

> [t]he complex subject matter of ABS [Access and Benefit-Sharing] its potentially far-reaching impact on uses of genetic resources and related information as well as the lack of detail in [the] Articles [. . .] have all combined to result in a very low level of domestic implementation by Contracting Parties to the CBD. By 2007, only 39 of the then 189 Contracting Parties had established domestic legislation or were in the process of doing so.
>
> (Buck and Hamilton 2011, 48)

Recalling the negotiations for the Nagoya Protocol in Japan in 2010, the protocol that has been meant to clarify the initial CBD, they point out that the key to really 'understanding' the real effects of the CBD and Nagoya Protocol is dependent on how national governments use their sovereign powers:

> The adoption of the Nagoya Protocol was a major achievement in international biodiversity policy making in 2010 [. . .]. Further international work preparing the entry into force of the Nagoya Protocol will be needed. However, most efforts over the coming years will need to be at domestic level, developing implementing rules to prepare ratification. In all Parties with well-developed or emerging research and development systems this will require significant awareness-raising with stakeholders from research and industry and will result in quite some discussions.
>
> (ibid., 60)

Most importantly, the national implementation has to take into account that access should take place on 'mutually agreed terms' and 'be subject to prior informed consent', conditions found in the original CBD and all subsequent treaties. However, other aspects of AnGRs can be regulated too, and some countries have already put in place requirements related to animal genetic material's import and export. The FAO's Technical Working Group on AnGR Access and Benefit Sharing Issues explained in a recent report, released in 2014, that

> [t]he sovereign right of states to determine access to genetic resources should not be confused with other categories of entitlement, such as the private ownership of an animal. A farmer's ownership of an animal may be conditioned by certain laws. For example, animal welfare legislation may regulate the handling, husbandry, and transport of the animal. Other laws may require the animal to be vaccinated against specific diseases, and so on. In a similar way, ABS measures may require that, even though an animal is the private property of a farmer or the collective property of a community, certain conditions

> (e.g., related to the need for 'prior informed consent') must be met before it can be provided to a third party for research and development.
>
> (Commission on Genetic Resources for Food and Agriculture 2014, Item 18)

Indeed, some of the countries have already exercised their sovereign rights. For example, China adopted a set of rules on AnGRs, or 'Measures of examination and approval of the entry and exit of animal genetic resources and the research in cooperation with foreign entities in their utilization', in 2008. These include a set of import and export rules, such as prohibition of 'the export of newly discovered and unverified' AnGRs in cooperation with 'any foreign institution or individual'. Also, any research into and use of AnGRs involving foreign collaborators requires permission from the Chinese authorities. As for flows in the other direction, South Africa requires a 'genetic impact assessment' before the import of new breeds. These studies need to be prepared by reputable South African animal scientists and submitted to the relevant authorities (Commission on Genetic Resources for Food Agriculture 2009, 34). National implementation related to the sovereign rights over genetic resources can be carried out in many ways, not only via regulation of access or benefit sharing but also in terms of use and impact, as the examples from China and South Africa demonstrate.

## Conclusion

The challenges are now located within the realm of national politics, the realm of the sovereign, where the *in situ* condition for genetic resources is turning animals into collections of nationally valuable animals, governed not under the previous ideals of global commons but via the logic of 'actual and potential' value, with innovated political re-categorisation of natural beings, and by national restrictions to access, use, and benefit sharing linked with AnGRs. In other words, we have moved from a world in which animals once were part of a seamless universal nature without boundaries to a world that is a collection of discrete 'national genetic landscapes'. These genetic landscapes are epistemologically fragmented spaces, where sovereignty, nature, and traditional models of ownership meet.

Over the course of the short history of AnGR conservation, the natural identities of farm animals have been refashioned. They have shifted from being objects of breeding to boost the productivity of individual animals and breeds to being objects that can be defined as actually or potentially valuable as nationally recognised genetic resources. The change in their identity is a creative output of the animal breeding and conservation sciences that have argued for the value of animals on the basis of scientific evidence as well as the global politics surrounding ownership rights over genetic resources considered valuable. AnGRs, including farm animals, are now as much political as they are scientific, as much 'cultural' as they are 'natural' in essence (see Table 5.1).

What becomes clear in looking at the key changes in the value system related to AnGRs is that AnGRs have become increasingly complex objects for breeders,

*Table 5.1* The historical changes in AnGR conceptualisation

| *ANIMAL GENETIC RESOURCES* | *PRE-CBD/ GPA/NAGOYA* | *POST-CBD/ GPA/NAGOYA* |
|---|---|---|
| **NATURE OF ANIMAL GENETIC RESOURCES** | • Natural breeds or functional genetic components<br>• Identification based on scientific definition and evaluation of conservation need | • Natural-cultural objects<br>• Identification based on the politically agreed *in-situ* condition<br>• Genetic material found within the geographically bounded territories of the nation-states, with the exception of politically innovated 'transboundary' category of animals |
| **ACCESS AND BENEFIT-SHARING** | • Private access, or club commons<br>• Freely usable for those who have physical access and local permission to use<br>• No enforced benefit-sharing system | • National sovereignty: access and use only under the rule of the sovereign party to the conventions<br>• Local political decision on access and benefit-sharing principles<br>• Binding and enforced benefit-sharing system |
| **VALUATION CRITERIA** | • Actual or potential value, no consensus in general | • Actual AND potential value<br>• Based on territorial *in-situ* condition and local political valuation of important national genetic materials |

scientists, and politicians alike, with no easy answers for the balancing of rights, responsibilities, and benefit sharing in the near future. While AnGRs have finally become a global issue with high political priority and prompting action, so have the political conditions under which the animals live become inherently global entanglement of science and politics, culture, and nature.

At the same time, the status of AnGRs that reside outside the CBD system – owned by either private companies or breeding societies before the entry into force of the CBD, in 1993 – is unclear. Although they are not objects of the CBD's articles, they might still be affected by and become targets of legal interventions – for instance, in the way in which China and South Africa have applied the sovereignty over genetic resources within their respective AnGR regulations. This makes the global system even more complicated, and most likely we will see a number of unforeseen challenging cases in the future.

The CBD, the Global Plan of Action, and the Nagoya Protocol present a global value system framing AnGRs in a way that is finally generating conservation action at national level. But on the global level, the system is more muddled than ever, calling for a great deal of conceptual, political, and legal analysis to bring more clarity to the current condition, which requires the generation of discrete genetic landscapes and marks AnGRs with their nationally correct *in situ* location as the political condition for their existence. Given the complex history of AnGRs as a global matter of concern, creating clarity to the present situation will not be easy.

At least three key elements need to be clarified with regard to AnGRs and the various claims laid to them if we are to move on in the global politics, in the creation and implementation of legal frameworks at national level, and in reflection on the true impact of the CBD and Nagoya Protocol.

1 What is the true scope of the CBD and Nagoya Protocol in terms of AnGR types? Are there types of AnGRs that remain totally unaffected by and reside outside the scope of the global conventions?
2 Do the signatory parties (nation-states) have prototypical reactions to – or at least broadly identifiable patterns in – the implementation of their sovereign powers over AnGRs?
3 If the signatory parties do exhibit identifiable patterns, a guiding typology of CBD and Nagoya Protocol implementation at national level would aid in making sense of how governments are adopting the global agreements (types of entitlement claims, access regulations, etc.) at large.

Addressing these three points would give a much richer and much more coherent overview of AnGRs' status in the post-CBD and post–Nagoya Protocol world than is currently available to the public.

Simultaneously, comparison with the material in the previous chapter, dealing with crop diversity rather than animals, reveals the idealism of such reform proposals. The complexity of the status of AnGRs as a global subject is identical to that of plant genetic resources when considered in terms of a fragmented global bio-scape. Certainly this is a dynamic that remains closely tied to more established notions of sovereignty, albeit in the performance and governing of collections of nationally valuable animals. Yet it is exactly this affirmation of 'the state' in the context of AnGRs that recalls what we called a 'zone of indecision', following Agamben (2005). Our perspective is not simply one that affirms the need for greater coherency between the performance of a stable internal order and the global attempts at balancing rights, responsibilities, and benefit sharing.

Again, we point to a suspension of the relationship between law and life as a permanent feature of the forces shaping the reality of our bio-infused present: what we have called metacodes are, then, about the reality of AnGRs as a global entanglement of science and politics, culture and nature, and about its suspension outside the context of the nation as a heavily disciplined set of concerns to be acted on and interacted with. Of course, this argument remains incomplete. The next chapter takes us away from genetic resources and global treaties, establishing the basis for a direct comparison in which our analysis of these subjects is considered alongside the rapid transformation of genetic engineering in the field of synthetic biology.

## Notes

1 This chapter is adapted from a previous version which was published in Tamminen, S. (2015). Changing Values of Farm Animal Genomic Resources. From Historical Breeds to the Nagoya Protocol. *Frontiers in Genetics*, 6, 279. doi:10.3389/fgene.2015.00279

2 The text of the treaty can be found online via www.cbd.int/ (accessed on 1.12.2016).
3 There are several ways to maximise diversity. Conserving a sum of isolates of pure inbred populations will allow saving rare genetic combinations adapted to specific environmental conditions but might result in losing overall diversity. Other options, such as maintaining a large outbred population resulting from crossbreeding, also would yield great diversity but do not usually fit the overall aim of conservation programmes. It is generally recognised today that a combination of *ex-situ* and *in-situ* measures applies complementary strategies.
4 The text of the UPOV treaties can be found via www.upov.int/portal/index.html.en (accessed on 1.12.2016).

## References


Agamben, G. (2005) *State of Exception*. Chicago: University of Chicago Press.

Boyazoglu, J. and Chupin, D. (1991) Editorial. In: *Animal Genetic Resources Information Bulletin*. Rome: Food and Agriculture Organization of the United Nations.

Buck, M. and Hamilton, C. (2011) The Nagoya Protocol on Access to Genetic Resources and the Fair and Equitable Sharing of Benefits Arising From Their Utilization to the Convention on Biological Diversity. *Reciel*, 20: 47–61.

Commission on Genetic Resources for Food and Agriculture. (2009) *The Use and Exchange of Animal Genetic Resources for Food and Agriculture*. Rome: FAO.

Commission on Genetic Resources for Food and Agriculture. (2014) *Report of the Eighth Session of the Intergovernmental Technical Working Group on Animal Genetic Resources for Food and Agriculture*. Report CGRFA/WG-angr-8/14/REPORT. Rome: FAO.

David, P.A. (2011) Breaking Anti-commons Constraints on Global Scientific Research: Some New Moves in 'Legal Jujitsu'. In: Uhlir, P.F. (ed.). *Designing the Microbial Research Commons: Proceedings of an International Symposium*. Washington, DC: National Academies Press.

de Haen, H. (1992) Opening Statement. In: Hodges, J. (ed.). *The Management of Global Animal Genetic Resources: Proceedings of an FAO Expert Consultation*. Rome: FAO.

Food and Agriculture Organization. (1990) Animal Genetic Resources: A Global Programme for Sustainable Development. *Animal Production and Health Paper No. 80. Proceedings of an FAO Expert Consultation*. Rome: FAO.

Food and Agriculture Organization. (1995) *Major Trends and Policies in Food and Agriculture*. Rome: FAO.

Food and Agriculture Organization. (1999) *The Global Strategy for the Management of Farm Animal Genetic Resources*. Rome: FAO.

Food and Agriculture Organization. (2007) *Global Plan of Action for Animal Genetic Resources and the Interlaken Declaration*. Rome: FAO, Commission on Genetic Resources for Food and Agriculture.

Hardin, G. (1968) Tragedy of the Commons. *Science*, 162(3859): 1243–1248.

Hiemstra, S.J.A.G., Drucker, M.W., et al. (2006) *Exchange, Use and Conservation of Animal Genetic Resources: Policy and Regulatory Options*. CGN Report 2006/06. Wageningen, Netherlands: Wageningen University and Research Centre; Centre for Genetic Resources, the Netherlands (CGN).

Hodges, J. (1984) Review of the FAO/UNEP Programme on Animal Genetic Resources Conservation and Management. In: *FAO Animal Production and Health Paper 44/1: Animal Genetic Resources Conservation by Management, Data Banks and Training* (2–9). Rome: FAO.

Hodges, J. (1990) Animal Genetic Resources. *Impact of Science on Society*, 40(2): 143–154.

Kloppenburg, J.R. (1988) *First the Seed: The Political Economy of Plant Biotechnology, 1492–2000*. Cambridge: Cambridge University Press.

Maijala, K. (1971) Need and Methods of Gene Conservation in Animal Breeding. *Annales de Génétique et de Sélection Animale*, 2: 403–415.

Mason, I.L. (1981) Cooperative Work by FAO and UNEP on the Conservation of Animal Genetic Resources. In: *FAO Animal Production and Health Paper 24: Animal Genetic Resources Conservation and Management. Proceedings of the FAO/UNEP Technical Consultation*. Rome: FAO.

National Academy of Science. (1993) Preface. In: Rice, B.J. (ed.). *Managing Global Livestock Genetic Resources Committee on Managing Global Genetic Resources: Agricultural Imperatives Board on Agriculture National Research Council*. Washington, DC: National Academy Press.

Phillips, R.W. (1981) The Identification, Conservation and Effective Use of Valuable Animal Genetic Resources. In: *FAO Animal Production and Health Paper 24: Animal Genetic Resources Conservation and Management: Proceedings of the FAO/UNEP Technical Consultation in 1980*. Rome: FAO.

Steane, D.E. (1992) Note on the FAO Expert Consultation on Management of Global Animal Genetic Resources Rome, 7–10 April 1992. *Animal Genetic Resource Information*, 9: 3–6.

Tvedt, M., Hiemstra, S.J., et al. (2007) *Legal Aspects of Exchange, Use and Conservation of Farm Animal Genetic Resources*. FNI Report 1/2007. Lysaker, Norway: FNI.

Wheeler, M.B. (2013) Transgenic Animals in Agriculture. *National Education Knowledge*, 4(1).

Wheeler, M.B., Monaco, E., et al. (2010) The Role of Existing and Emerging Biotechnologies for Livestock Production: Toward Holism. *Acta Scientiae Veterinariae*, 38: 463–484.

Wilmut, I., Haley, C.S., and Woolliams, J.A. (1992) Impact of Biotechnology on Animal Breeding. *Animal Reproduction Science*, 28: 149–162.

# 6 Recoding synthetic lifeↄ

## From openness to (free as in) freedom

> It seems to me that in Cynicism, in Cynic practice, the requirement of an extremely distinctive form of life – with very characteristic, well defined rules, conditions, or modes – is strongly connected to the principle of truth-telling, of truth-telling without shame or fear, of unrestricted and courageous truth-telling, of truth-telling which pushes its courage and boldness to the point that it becomes intolerable insolence.
>
> Foucault, the lectures on 'the Courage of the Truth', 2011 (1983/4): 165

### Introduction

Some years ago, we were introduced to Synthia, which is the nickname for a synthetic genome that was genetically engineered and patented by the J. Craig Venter Institute. While most attention was directed to the suggestion that an artificial life form had been created, the accompanying press release referred to the experiment as a step towards the ability to 'activate' and 'boot up' cells.[1] Here, we take this informatics metaphor as a starting point for examination of the field of synthetic biology, asking what a counterpart to the organism nicknamed Synthia might look like. Let us call her Cynthia, who leads an 'Open Source' kind of life and who might, if she so chooses, try to grow up to be free from influence, like the ancient Cynics, who renounced wealth, reputation, and power.

Synthia was announced in 2007 with this headline: 'Goodbye Dolly . . . Hello Synthia!' Hence, her personality was immediately cast in the same light as the controversies surrounding the cloning of animals, and, like the creation of Dolly the sheep, Venter's experiments were challenged on ethical, social, and environmental grounds (ETC Group 2007a, 2007b; see also Franklin 2007). The ETC Group, the civil society organisation that came up with the nickname, pointed out that there were patents on the sequenced and the synthesised version of the genome. To them the outcome of the experiment pointed in a familiar direction: one day soon, the strings of synthesised DNA that had been re-assembled in a living cell by Venter and colleagues could be put to work to create whatever synthetic compound might be in demand on the world market – plastics, chemicals, oil, and the like.

This chapter, in turn, asks what kind of alternative could take shape in synthetic biology. How could a counterpart to Synthia be imagined, again personified as a living embodiment of synthetic biology but simultaneously being a representative of the potential for another type of relationship to those who will have to live and work with the synthetic life forms under construction? We call her Cynthia – but not to brand her with crudeness, ignorance, and the lack of culture of the stereotypical cynic. She might be insolent in regard of conventional boundaries between life and technology, but she is simultaneously an 'extremely distinctive form of life' with a tendency towards 'truth-telling' and towards the search for what 'is indispensable to human life or which constitute[s] its most elementary, rudimentary essence' (Foucault 2011, 171).

Before we introduce Cynthia properly, some observations are needed on the ownership of knowledge in the life sciences. The first section outlines how the ownership of knowledge is being re-established on the foundation of the dream of a world wherein sophisticated approaches to knowledge about life and genetic engineering are situated outside the complex rules and controversies for which the life sciences are notorious. To set the stage for the comparison requires a further intensification of our theoretical analysis, applied directly to 'how' the convergence with informatics has changed the status of the patenting of DNA. Keeping the patent in its place requires a reconfiguration of practises, institutions, and forms of capitalisation that cut across the process of extracting information from living materials and its eventual re-materialisation as bioproducts – as food, medicine, energy, and so forth.

In the second section, we get to know Cynthia 'in person'. On the surface we can reach out by discussing her affinity for the activities of the design community associated with the BioBricks Foundation (BBF), which has created a registry with thousands of biological parts that are accessible over the Internet. Our objective, however, is to see her for who she really is, which means going beyond the rhetorical difference between patenting and Open Source. We compare the minimal genome project (aka Synthia) to the minimal cell project (aka Cynthia), which involves many of the same protagonists as the BBF. This is also why there is a 'copyleft' symbol in the title of the chapter. This 'reverse copyright' sign points to a comparison between the ethics of Open Source in synthetic biology and the response in informatics to the accommodation of source code by the free software and Open Source movements.

## Recoding life in common

### *Extracting information from oceanic bodies*

> Ah! sir, live – live in the bosom of the waters! There only is independence! There I recognise no masters! There I am free!' Captain Nemo stopped at these last words, regretting perhaps that he had spoken so much. But I had guessed that, whatever the motive which had forced him to seek independence under the sea, it had left

> him still a man, that his heart still beat for the sufferings of humanity, and that his immense charity was for oppressed races as well as individuals.
>
> – Jules Verne, *Twenty Thousand Leagues under the Sea*, (1992, written in 1870, Chapter 10)

More than a decade ago, Craig Venter went fishing, sampling microbial life from his yacht the *Sorcerer II* as he travelled the world. His expeditions continued from 2004 until 2010, but it was the initial journey that got most publicity, reaching a global audience through Discovery's epic television specials. These showed Venter as a heroic individual driving forward the frontiers of science. Simultaneously Venter's media-friendly genetic circumnavigation of the oceans on his yacht caused controversy through its symbolism, consciously calling to mind Charles Darwin's explorations as well as the journeys of marine adventurers such as Jacques Cousteau.

It is in this regard that the citation recalls Captain Nemo, the outlaw submarine captain from the novel *Twenty Thousand Leagues under the Sea*. Unlike the voyages of Darwin and Cousteau, a comparison with Captain Nemo pinpoints the twisted manner wherein Venter became an unlikely champion of the ocean as a commons. The global audience of the expedition became witnesses of 'Venter's captain Nemo fantasy about the free seas'. What the citation shows is a fantasy wherein the ocean is a place without masters, of freedom and independence from society and its rules. Nemo searches for independence, and so did Venter and his supporters when they confounded their critics by making the data extracted from the oceanic samples they collected available in the public domain.

What was confusing was that Venter was expected, and not without cause, to behave in a way that had made him into a global celebrity as the CEO of Celara, which is the company that attempted to patent hundreds of genes in direct competition with the public universities and institutes that had started the human genome project. The latter even attempted to make such patenting more difficult by releasing data into the public domain as quickly as they could and working with a higher accuracy. Unsurprisingly the countries with jurisdiction over the waters that his yacht came into contact with reacted on the assumption that he was after their valuable genetic resources. They maintained that such access requires benefit-sharing in line with the Convention on BioDiversity (CBD) that had been in place for nearly a decade (as discussed in Chapters 4 and 5). Who was to know what it was worth? Perhaps the data derived from the unsequenced microorganisms out of the ocean might – after being recomposed using bioinformatic technologies – turn out to be highly efficient in transforming sunlight into energy as Venter had announced (see Delfanti et al. 2009; Pottage 2006).

As Venter put it in a later interview:

> [W]e're sailing across the open ocean in international waters and there's this current moving across the Pacific at 1 knot. So there are microbes in that

> current that move from open ocean into the 200-mile limit of French Polynesia, and suddenly the French call that French genetic heritage. Right? And they want to own it and capitalize on it. It takes months of paperwork to take 200 litres of seawater now from the open ocean. Before we published our paper nobody cared, because nobody assumed anything was there. So I think it's quite comical that we're called pirates for describing the data and making it available for the world.
>
> (quoted by Rimmer 2009, 171)

The citation shows that Venter realises that the oceans are hardly a common heritage any longer. Moreover, those areas where they are still considered a global commons are not on the outside of international treaties or generally accepted scientific procedures that marine biologists have to abide by. Nevertheless, Venter's expedition easily deflected any accusation of 'biopiracy' of genetic materials. It is, after all, difficult to place a claim on something valuable when it is being given away for free.

This time, however, Venter did not attempt to patent the data, and neither was he charging for access to the information extracted from the samples. Yet the analogy with Captain Nemo still breaks down. Even when dealing with the oceans, the territories of this earth have been claimed: Robinson Crusoe does not get to live on the island, Frankenstein's monster could not stay in the Arctic, John Savage has left the Savage Reservation of the brave new world, and there is no ocean where synthetic biologists can work in perfect isolation. Each of these fictions is an example of Rousseau's state of nature whose inhabitants are independent and free at a distance from the rules of the world's civilisation, in this case featuring the oceans as the exterior of the national world on the outside of sovereign jurisdictions. The same applies to Venter's contribution to oceanography: his concern was not the global commons, territorial or environmental; he was mapping the oceans' genetic material in order to know the interior of the natural world. What he was after was to open up the natural world even further, this time around through its transformation into an informational horizon of infinite complexity, making visible for an instant the establishment of a frontier area that is not within the reach of patentability and state authority.

The comparison is, therefore, a different one. Rather than stopping with Captain Nemo or with Rousseau's state of nature, we have to go further back in time: to Hugo Grotius, the author of *The Freedom of the Seas* (1609). Grotius wrote on behalf of the Dutch companies who were opening up the passage to the East under the control of the Portuguese. The basic notion that Grotius defended was well known at the time, which is that there was a difference between states' jurisdiction over neighbouring waters and with land. Following a number of commercial conflicts in the Indies, Grotius began to defend the rights of the Dutch to seek trade and to choose whether to make agreements with its inhabitants or take by violence what they wished. What he argued was that no individual could own the oceans and therefore it is impossible for states to have comprehensive jurisdictions over the high seas.

Even though geneticists on highly publicised sailing expeditions might no longer need the compass and sextant of Grotius's times, the dynamic of opening up a new terrain by claiming rights of passage and usage is the same. When Grotius granted natural rights to civilised/colonial men, this implies that they could make use of the natural world beyond their societies. This argument resurfaces with every natural philosopher with a state-of-nature theory afterwards until Rousseau: every right held by a state must be derived from an individual, and if an individual could not own something, then this right could not be transferred to a state of society (Tuck 2003, 82–92). Of course, Grotius was doing exactly what Rousseau later denounced as typical of how all who are 'reasoning on the state of nature, always import into it ideas gathered in a state of society' (Harvey 1996, 163; see Rousseau 1993, 65). For example, Grotius's state of society mirrored the interests of Dutch naval companies as the main beneficiaries of his definition of the natural right to property and the right of self-defence. These companies, when going beyond the jurisdictions of 'civilised' states, were granted the 'most far-reaching set of rights to make war which were available in the contemporary repertoire' (Tuck 2003, 108).

A generation later, Grotius's rights to violence were integral to the work of Thomas Hobbes when he posited that it is only the sovereign who can guarantee any right to property. This, however, suggests a very different comparison. A Hobbesian world view, extended to Venter's fishing trip, is one wherein only states can guarantee the recognition of patents and compensations. In this case genetic resources might be within a sovereign jurisdiction or some states might be able to enforce their preferred ideas about intellectual property. To push the analogy, it is through the assertion of an oceanic genome to be catalogued and taken back home that Venter calls back to Grotius's and Hobbes's lack of knowledge of the inhabitants of the newly discovered territories. Neither Hobbes nor Grotius considered their state of nature a fiction; the former saw nothing but violence in the behaviour of those he considered to live outside civilised society, and Grotius describes the Americas and the Indies as being in a 'primitive state [. . .] exemplified in the community of property arising from extreme simplicity' (Grotius 1993, Book II.II.2 as quoted by Arneil 1996, 49).

The comparison is not confined to sovereign claims on oceans or biodiversity. After all, there are contemporary candidates that can be seen as the inhabitants of an equivalent to the early-modern version of the natural world. The comparison is not about how people today continue to be reported and depicted in ways that resemble the ignorance of Hobbes and Grotius's references to savages. Rather, their ignorance of the natural world can be followed to how the living entities sampled by Venter and his team were described without attention to environmental context – seasons, temperatures, salinity, and so on. What this does is assert an overly simplistic oceanic genome as an outside where the rules can be rewritten to establish the same familiar claims of property and authority.

Sometimes states plant flags, whether on genetic resources or where the melting of ice caps might leave oil in its wake. At other times, the map still has many blank spaces, whether it is in regard of the contemporary exploration of the oceanic genome or the Dutch incursion into the colonial map of the world as the Portuguese saw it. Territory shaping and map remaking invariably implies a return of state-of-nature theories, and following the suggestion of natural foundations, we arrive at contemporary notions of social contracts between science and society. Here a globalised frontier area is shaped through the collection of samples and through extraction of data to be made freely available. The question is who ends up featuring as the inhabitants of how this contemporary equivalent to the state-of-nature theories, taking their places alongside the natural foundations that are being re-imagined. With Venter it is the genetic wealth of the oceans, transferred from the outside of the jurisdictions of society over territory and resources to its inside, as freely available data. Along similar lines we turn to the multidimensional representations and translations of genetic materials into complex biological entities that are at some unspecific point in the future to be re-materialised in the name of progress and all that's good about contemporary societies.

To put it differently, it is through forms of life that sovereignty comes into being (Tamminen and Brown 2011, 2). However, this leaves open 'how' exactly this coming to being occurs. What a comparison with Grotius's view of the oceans demonstrates is that the controversy is not merely about a frontier in between nature and society or a new type of 'border crossing' for national claims on novel types of biological resources or for the free sampling of the oceans. Either position is constitutive of a state of society, going along with the bioinformatic returns to the state of nature as a foundation myth. Biology's return to the ocean as the mythical origins of all life, and of natural history as a field of knowledge, is mirrored in the life sciences as a setting that shows how sovereignty is 'maximised' through the mobilisation of corporealities.

### *Exchanging DNA as information*

Unlike Rousseau's state-of-nature theory and the fantasy of free seas, Grotius justified the opening up of the natural world as a necessary condition for establishing ownership and authority. The same applies to the life sciences in its relationship to the interior of the natural world. What lies between the life sciences and the ownership of knowledge is the extraction and recoding of information extracted from living materials, which has to be actively mobilised and redirected in order to consolidate the patenting of DNA.

The key step has already been taken. The footprint that Crusoe found in the sand (Chapter 4) implied that the establishment of the patenting of DNA was a decisive event that opened the potential for numerous biological terrains to be opened up to private investments. The patent as an institution is expected to function as a reward for inventions that grants exclusive rights, enabling temporary monopolies that make it possible to recover the necessary investments. Patenting was developed in response to advances in organic chemistry and the rise of related industries in the late nineteenth century, including the appearance of research laboratories in corporations (Dutfield 2003). With the life sciences, however, this means that its introduction is based on an extension of the criteria for the patenting of chemical compounds. It is within this context that the patenting of DNA began, typically traced to the early 1980s US Supreme Court ruling establishing that DNA could be a 'technical subject'. The implication of this decision was that legally speaking, certain types of DNA were designated as a 'composition of matter' and a 'product of ingenuity' rather than a 'manifestation of nature' (Parry 2004, 85; Calvert and Joly 2011).

Genes, like chemicals, were suddenly declared to be a product of human ingenuity rather than found in nature, an invention rather than a discovery as they had previously been (van Dooren 2007). The same applies to the numerous other verdicts around the world that expanded the scope of patenting, thereby affirming the status of DNA as an invention and hence as 'technical' rather than biological or natural. For example, it became possible to patent the isolation of DNA in a purified form – as sequenced strings of DNA that can be considered useful and valuable enough to be subject to patent rights. In these cases the main condition is that the DNA was not purified before, and it makes no difference where the DNA was taken or that it might have already existed for some time (see Carolan 2010).

Simultaneously, the patenting of DNA should not be understood as a mere continuation of what happened in the early 1980s. Consider, for example, the US Supreme Court's invalidation of the patents held by Myriad Genetics for two genes with mutations that cause cancer (BRCA1 and BRCA2) in 2013. The company had patented the two genes as diagnostic tools for the testing of breast cancer. Therefore Myriad would be able to ask higher prices for its tests and restrict access to medical information obtained through those tests. The case got widespread publicity because of the suggestion that there may be stricter criteria for patenting in the US. What is significant, however, is not only whether the legal criteria for patenting are changing but that diagnostic tests, such as Myriad's, are quintessentially about the invention and usage of informatic artefacts and related practises. The original ruling by a district court judge had invalidated the patents, with the observation that the DNA involved was known through its 'information content, its conveyance of the genetic code'. Similarly, the final ruling by the Supreme Court in 2013 explains that Myriad's claims were not expressed in terms of chemical composition and did not create or alter the genetic information encoded in the genes or the genetic structure of the DNA (in the case *Association for Molecular Pathology v. Myriad Genetics, Inc.*).

What matters here is not the invalidation of patents or the limitation of their scope out of concern over whether patents are the most appropriate means to facilitate the distribution of information. Either result simply underscores a premise shared by the advocates of patenting and the critics who advocate for openness (Lawrence Lessing, James Boyle, Yochai Benkler, etc.). The shared objective is the acceleration of the development and introduction of novel technologies (see Hilgartner 2012, 192). After all, the patented invention has to be described in the application, and the information is released to the public when the grant expires, thereby guaranteeing its reproducibility.

The case of these two genes illustrates how the biology requires that more legal work to be done for DNA to be turned into an object that can be patented. It is only under certain conditions and with considerable difficulty that informatic practises can be identified with DNA that has stable chemical properties (Calvert and Joly 2011; Caulfield 2011). Hence, there will be patents and there will be patent controversies to be settled, as more legal work will be needed to separate DNA as a composition of matter from DNA as an informatic entity. Simultaneously, however, it is a biological object that requires cheap, efficient, and high-quality access to interactive databases, software, and hardware, to facilitate labour in the form of downloading, copying, and searching for information the world over via the Internet.

Many examples show that property rules, institutions, and practises are being reconfigured in a strategic relation with the patenting of DNA. For example purified DNA as a 'composition of matter' (by analogy to patenting in chemistry) is simultaneously an informatic entity. It might rely on databases with pre-programmed homology searches of databases that make information available in the public domain. It might simultaneously have its DNA patented while copyrights are granted automatically for the programming source code. There are many possibilities, such as software patents on processes that make computers run faster, interfaces for better interaction, or other types of improvements in efficiency.

Similarly, we can look at the specially developed series of supercomputers called Blue Gene. Such supercomputers are necessary for any type of model that is based on tera- and peta-bytes of information, and this includes models that requires its processing power for the simulation of the complexity of the interactions that take place in the context of cells. For example, the Blue Brain Project is one of several supercomputers in which Blue Gene is used: its objective is 'to reverse engineer the human brain and recreate it at the cellular level inside a computer simulation' with the aim of developing treatments for brain disease.[2] Furthermore, Blue Gene is Linux based: Linux has long been the key working example of open source as 'a collective project that has been shared and worked on freely' (Berry 2004, 80). This demonstrates the increasing likelihood of information being released by using the licences that remove the possibility of restricting the usage of information, whether studying it, changing it, or redistributing it.

The choice between the two models is increasingly procedural, with the development models (open vs. closed) seen as applicable in different circumstances. Openness is the option that guarantees the ability to modify programmes and models to allow for an increasingly wide variety of researchers with highly specialised knowledge and different levels of commitment to work together. The examples are numerous; there are many 'open bio' versions of programming languages like BioJava, BioPerl, BioSPICE, and so forth. Each has its own voluntary or public science support communities like the Open Bioinformatics Foundation. Similarly, there are different kinds of databases that use Creative Commons legal tools, such as the Material Transfer Agreement, or regular open licencing, for example in a protein database called the 'neurocommons' and the 'hapmap'.[3]

However, there is a further step to take, engaging directly with how the tension between patenting and openness has taken hold over life as a technological creation in synthetic biology. Again this is about 'life in common', although the issue here is not the extraction of information from the environmental commons (e.g., the ocean) but a geopolitical claim on a natural world wherein living materials are being translated to digital formats. This sets the stage to get to know a new biological form of existence – one that might be 'radically other' in how it is engineered and as a practical relationship to nature that informs what social life ought to be.

## BioBricks™ or synthetic life🄯

> The process of extending engineering principles to biology and applying business principles to genome engineering were logical and indeed inevitable developments.
>
> – George Church in *Regenesis: How Synthetic Biology Will Reinvent Nature and Ourselves*, 2012, 75

### *BioBricks™*

Is the extension of engineering and business principles to synthetic biology **inevitable**, as the opening citation suggests? Perhaps; after all, the costs of gene synthesis and editing might very well continue to fall and mastery of these techniques will increase, possibly even exponentially as in the realm of computer processing. Yet application of business and engineering principles is not nearly as inevitable as suggested when followed in a different direction. Specifically the direction to go is the one indicated in the sign ('🄯') in the heading of this section as well as the chapter's title.

The sign for a 'reverse copyright' has a distinctive relationship to the extension of intellectual property to the life sciences. This is the principle and movement of copylefting, popularised by the Free Software Foundation (FSF), the Open Source movement and more recently adapted by the Creative Commons and various other organisations and sectors. Its origin is the moment of copyright being extended from the texts of novels and newspaper articles to the ones and zeros of source code and information in datasets. Unlike patents, copyrights are automatically granted to authors, and this is also true for a copyleft. The difference is that a copyleft implies that the right to restrict the distribution of data and modification therefore is given up. Accordingly BioPerl© and BioPerl🄯 might refer to the same programming language that allows life scientist to write their own programmes and ask novel biological questions, but only the latter guarantees that the software can be modified and developed further by any of its users. This is a condition of the copyleft: other should be able to use, study, copy, modify, and redistribute the information in any of its forms.

The life sciences are one of many sectors wherein copylefting has become a regular occurrence, accompanying the usage of software and databases. Often there is little to indicate that such licences are in effect, since copylefting has become nearly taken for granted as a common-sense means of guaranteeing that information remains accessible and of facilitating sharing and collaboration on the gigabytes of data as well as the hardware and software that run models and simulations that are integral to interrogate the biological in its complexity and interactivity (Deibel 2013). Accordingly, Venter's publicly released body of data on the oceanic genome and various bioinformatic tools co-exist with copylefting as a principle that is at least as inevitable as in the application of business and engineering principles that George Church mentions in the opening citation. Furthermore, the copyleft principle can also be found in experimental settings where sophisticated approaches to genetic engineering are identified with the values of openness, collaboration, and the public domain.

The principal example is found, again, in the field of synthetic biology. In particular, the BBF has in a short time come to symbolise 'the pragmatic yet speculative ethos of the growing discipline of synthetic biology' (see Mackenzie et al. 2013, 709). The BBF revolves around a registry of thousands of biological parts that are accessible over the Internet, which is considered the principle platform for the coordination and acceleration of the circulation of knowledge and the development of novel technologies. There is an open licence, the BioBricks™ Public Agreement, which has as its aim that anyone can 'make their standardised biological parts free for others to use'. This is what sets the BBF apart from most genetic engineering efforts. It is a front runner with its claim that genetic engineering is an area of science that is inclusive, transparent, and open. Such values are what the BBF's slogan is premised on: 'the engineering of biology can be an ethical approach that benefits all people and the planet'.[4] The website of the BBF raises the expectation that synthetic biology will result in a genuine alternative to a closed model of genetic innovation. The insistence on such an ethical approach, however, remains largely examined in regard of the meaning of 'free' and 'open' in the context of the 'inevitable' extension of engineering and business principles.

### *The business of synthesising DNA*

For some time now, many companies have been doing DNA synthesis, which is something that spares life scientists in various fields of specialisations the time-consuming effort of doing DNA synthesis themselves and makes increasingly complicated projects possible (Bügl et al. 2007). One notable example is the case of Codon Devices, which was a start-up company founded in 2004 with venture capital that specialised in the delivery of synthesised DNA and the design of related applications. Crucially, Codon went out of business in 2009 because there was little profit to be had in the synthesis of DNA, which is becoming cheaper. Also Codon was not successful in finding customers for the design of applications. This company was home to many of the key proponents of the BBF, and its failure is significant as an indication of the competition among companies doing DNA sequencing and synthesis. This is where the BioBricks Foundation is important: genetic engineering as a field of activity ranges from the dramatic lowering of costs of the translation of DNA into informatic formats to the much more difficult re-materialisation of this information through DNA synthesis.

The function of the BioBricks™ registry is to guarantee access in the sense of allowing for any 'useful purpose' that synthetic biologists might have come up with. A frequently used basic definition of synthetic biology expresses it in these terms: synthetic biology is a field that aims to design 'biological parts, devices and systems for useful purposes' (see Rathenau 2006). This matches a way in which the distinctiveness of synthetic biology is frequently characterised on the basis of how the field has shifted from the behaviour of the whole genome and the understanding of biological systems as a whole to the 'capacities for editing the interaction of the parts' (Brent 2004, 1213–1214; see also Knight 2005). These parts have been engineered 'to meet specified design or performance requirements'. Accordingly 'genetically encoded objects' perform biological functions and can be connected as 'functional inputs and outputs' of different parts, which are sometimes referred to as biological equivalents of sensors, logic gates, and actuators. Each new sequence gives researchers reason to consider 'additional natural genetic parts', such as 'protein coding sequences, regulatory elements for gene expression and signalling and other functional genetic elements' (Canton et al. 2008, 787). The ethics of the BBF are derived from the principle of free distribution, or a licencing scheme that aims to prevent these biological parts from being patented, since that patenting would undermine any sense of a 'freedom to operate' for synthetic biology (Henkel and Maurer 2007; see also Rai and Boyle 2007). Similarly, there is a strong emphasis on inclusion, which is exemplified by an annual competition for students called International Genetically Engineered Machine, or iGEM. Its size has grown vastly, to the point where teams are formed all over the world and the Web site describes the aim as being the development of an 'open community and collaboration'. Examples of what the students contribute are 'biological devices to make cells blink' and projects ranging from wintergreen-scented bacteria to the development of an arsenic biosensor to screen drinking water (see also Baker et al. 2006).[5]

The suggestion is that genetic engineering is becoming something that is easy to do. The most telling example is the teams of students coming up with novel applications that behave in accordance with protocol and that can be produced independently as biological parts that are then available for use in other projects. Conceptually, this has been called 'modularity', a term that Yochai Benkler has explained as 'the number of people who can, in principle, participate in a project', which 'is inversely related to the size of the smallest scale contribution necessary to produce a usable module' (Benkler 2006, 101). From this standpoint the BBF and iGEM show that BioBricks™ design is modular enough to allow widely distributed and loosely connected individuals to cooperate with each other. More generally, the suggestion is then that synthetic biology should be conceived of in terms of a new type of economics that revolves around 'sharing' and 'peer production' (see Benkler 2006; Kelty 2004, 2012). The problem, however, is the step from the modularity in bio-informatics, which is highly modular, and the design of functions, to a growing community of open, creative, motivated, and skilled contributors that has taken shape.

From a certain perspective this is in line with how DNA synthesis can be considered highly modular, in that the technique relies on software to facilitate a 'multiplicity of techniques coordinated on an elevated surface (the screen)' (Mackenzie 2010, 189; see also Newman 2012; Deibel 2009). As Andrew Mackenzie has explained with regard to DNA 2.0, another company involved in design of synthetic DNA, the improvement and optimisation of its properties revolves around digital sequences that are:

> flanked by various genetic promote, regulator, start and stop sequences. The actual codons that comprise the synthetic DNA construct can then be 'optimised' by the software to confer optimum expression of the target protein construct in the chosen host. Once a design has been assembled using these components, Gene Designer can check it for errors, optimize it in various ways (for instance, for expression in different target organisms), check if for completeness and then provide an estimate of the cost of synthesis.
>
> (Mackenzie 2010, 188, 189)

The suggestively named application ('Gene Designer') invokes the rhetoric about synthetic life being entirely artificial within a business model that is an example of how the expression of DNA in digital or electronic forms can be 'acted upon and interacted with in ways that would not otherwise be possible' (Parry 2004, 65; see also Pottage 2006). Yet the 'Lego-like' characteristic of BioBricks™ appears exaggerated when its immateriality is accentuated. This is further illustrated by closer examination of the citation that began this book, from a company called 'Thermo Fisher'.

> The new Invitrogen™ GeneArt™ CRISPR Search and Design Tool allows you to quickly search our database of >600,000 predesigned CRISPR guide RNAs (gRNAs) targeting human and mouse genes or analyse your sequence of interest for de novo gRNA designs using our proprietary algorithms. Up to 25 gRNA sequences per gene are provided with recommendations based on potential off-target effects for each CRISPR sequence. Once you've selected the optimal gRNA designs, you may purchase your gRNAs and other recommended products for genome editing directly from the Web tool.
>
> (see www.thermofisher.com/ last checked in February 2016)

The citation refers to Thermo Fischer's database, proprietary algorithms, and genome editing directly from the Web-based tool. This illustrates how the capturing of value from the representation of DNA in a more 'purely informational form' – as data or images – will be 'standing in for particular materials resources henceforth to be absent' (Parry 2004, 65). What is absent is the kind of materiality of scarce and valuable resources that disappear when priority is given to informatic ways of thinking about life and nature, which no longer represent the extraction of information from living materials.

Even more than with Venter's micro-organisms, there is no need to engage in bio-prospecting or to negotiate 'access and benefit-sharing agreements' because the information is freely available in nearly infinite quantities (see Hayden 2003a, 2003b; Hoare and Tarasofsky 2007). What this suggests is that the activities of the BBF are similarly detached from any suggestion of life as having value (as going extinct, sacred, economically scarce, etc.). This is only the case, however, with regard to biological and cultural diversity, for recently the pendulum has swung back from a discourse that prioritises informatics towards chemistry. Several of the start-up companies have not been successful, showing, just as Codon did, that there is considerable competition between companies that specialise in efficient access information and its speedy delivery as synthesised DNA. The relationship to chemical companies suggests a return to a model organised around the patenting of DNA and established market models. Work to mobilise specialised knowledge and involvement in the design of applications is no longer a reflection of the expectations that once were attached to the informatic basis of the genetic techniques. Instead the work is centred on the ability to add value to the nexus of (petro-)chemical industries as well as health and agriculture.

It remains to be seen whether these companies can deliver products that are able to compete with established industries, especially in light of raising these expectations, which makes it increasingly difficult to identify 'where it is in biology that value resides' (see Helmreich 2008). However, the emphasis on biochemistry implies that priority is given to exclusive ownership. On the one hand, we have seen how information has been extracted from the global commons (Venter/the ocean) and turned into a commons of freely available information and related techniques, some of which are found in synthetic biology. On the other hand, we have seen that the affirmation of such a boundary area was a strategic necessity that sits between informatic ways of thinking about life and the established rules of property, business, and commodity chains, whether as medicinal test, a genetically altered crop, a biofuel, or otherwise.

## *Minimal species of life™*

> For life, like a machine, cannot be understood simply by studying it and its parts; life, to be understood, must also be put together from its parts.
>
> (see Church and Regis 2012)

There is nothing 'inevitable' about how markets and technologies come together in synthetic biology, either in terms of markets or as an approach to the engineering of biology. What is inevitable, however, is the 'mistake' of interpreting 'a form of life' that is dynamic and multifaceted as 'a life form' (Mackenzie 2013, 712).

Let us consider two examples. The first is a project called Open Worm. Its aim is similar to the other types of software projects discussed; it seeks contributors to programme a complicated piece of Open Source software to simulate the behaviour of a microscopic roundworm with a low amount of cells. What is significant is the project's tag line, which equates its efforts with: 'building the first digital life form. Open Source'. This project makes the mistake that is mentioned earlier. Its view of the 'building of life' is primarily discursive, presenting a digital model of a form of life as a life form, thereby mixing naturalistic and informatic language. The second example is found in a new start-up associated with the

Danish pharmaceutical company Novo Nordisk called BioSynthia, which markets itself as a 'catalyst for change' with promises to produce synthetic counterparts to chemical synthesis in the name of a 'greener tomorrow'. Here, naturalism is prioritised, depicting a green alliance between synthetic biology and the petrochemical industry as the agents of inevitable change that are realising the sustainable future. This time, however, the mistake is made by how its name refers to synthetic biology and the minimal genome project.

Synthia as a life form is the embodiment of several forms of life that are the object of genetic engineering in a series of experiments. The first of these involved cloned smaller stretches of DNA that were re-assembled into a bacterial chromosome. To do this the repair mechanism of yeast was used to create the largest synthetic bacterium ever made inside a cell. The procedure revolves around systematically 'knocking out' the protein-encoding genes to find the smallest set of genetic material necessary for the genome to survive under controlled laboratory conditions. Venter and his colleagues chose a bacterium with the smallest known genome (*Mycoplasma Genitalium*) and showed that some 20% turned out to be 'dispensable'. Subsequently they synthesised the entire genome (*Mycoplasma Genitalium* JCVI-1.0). The third experiment was the one that captured most media attention. Using two different types of bacterial species, they transferred one set of genomes into recipient cells, which thereby 'turned themselves into members of the first' (Church et al. 2012). Finally, they assembled another genome (*M. mycoides* JCVI-syn1.0), which was transplanted into a recipient cell with the same result (ibid.).

In turn, we submit Cynthia as the nickname for a life form alongside Synthia, Dolly the sheep, 'terminator seeds' and many other examples of 'species of life™', which are alive and invented (Deibel 2009; see also Haraway 1997; Franklin 2007; Van Dooren 2007). These are her relatives, with Cynthia as a new member of the family and another living embodiment of the various ways wherein possible alternatives are suggested by, for instance, associating digital models of life forms with open source or by combining experiments in synthetic biology with the greening of chemical industries. What is particular about her? Crucially, she seeks to remind us that there are different types of principles on which the design process could be based. She is not the materialistic type, nor does she believe that genetic engineering should be something easy and fun to do – something that is 'modular' enough to involve others. This is her most redeeming quality: she advocates openness and really does not think that her work belongs only to her. In other words, she is committed to lead an 'open source' kind of life.

As discussed, modularity in the life sciences is exaggerated. It might apply to bio-informatics but much less so to communities interested in genetic engineering. Therefore it is not so certain that she is really that different from her sister Synthia, who had few choices in life because of how debts and investments were made to bring her into this world. We don't know yet: the story so far is rhetorical: we've only made a quick attempt to give life to Cynthia by modelling her personality on the designs of the BBF. What we have to do is to bring into the discussion the kinds of experimentation that brought the two sisters into existence as synthetic life forms. After all, Synthia is the living embodiment of the 'minimum genome project', which is how Venter and colleagues attempted to identify the essential genes of a minimal bacterium. In turn, we can get to know Cynthia better as her personality forms at the intersection of the life sciences and Open Source in informatics. Specifically, there is an attempt to build a minimal living cell 'that we create ourselves, from the ground up' (see Church et al. 2012).

Accordingly, Cynthia's creators were critical of these experiments, pointing to how only a tiny portion of the cell is synthetic: a synthetic genome that depends on the recipient cell's natural and native apparatus for its expression. In a counterproposal they explained that the minimal genome continues to rely on complex combinations of gene expressions and that removing these one by one leaves many 'false essentials', unnecessary elements that can be removed if one had taken overlap and interaction into account. Also many 'essentials' have unknown functions that leave much of the complexity of a bacterium or cell intact, and they suggest that this means that '[b]uilding a machine from mysterious parts can only create a mysterious machine' (Forster and Church 2006a, 1–2). As Church explains, if the first level is a minimal genome, then the second level is a living entity put together from its constituent parts and 'composed of the fewest components that can jointly carry out all the normal processes of life' (ibid.). Accordingly the synthetic genome is one of many biochemical subsystems that are needed to 'encode a near minimal, self-replicating system dependent only on small molecule substrates' (Forster and Church 2006b, 4).

Unlike *Synthia*, who came into existence through experimentation, this 'minimal cell project' has not yet been realised. Nevertheless it provides a further glimpse, alongside the BBF and the 'synbio' business model, of *Cynthia*'s outlook on life. She will set out to minimise whatever is non-synthetic in her life. She actively questions how synthetic she really is, and, in all likelihood, she will be good whatever she puts her mind to. Furthermore, she seeks to collaborate, which reflects the BBF's position as one of the preconditions for the proposal of biochemically defining each of the functional processes taking place in a cell.

Yet her ambition is likely to blind her to the way in which she represses her (biological) uncertainty, variation, and fragility whenever it is out of her control. After all, even the minimal cell project will not be as synthetic as her creators imagine it to be or to become. They know that inevitably some uncharacterised components will be ignored or removed that turn out to be needed under different circumstances. Similarly biological variation continues: the biochemical circuitry of the bacterium will grow, replicate, and interact with the host cell, while pre-existing or synthetic pathways will bind together in unexpected ways. Even though there is the suggestion that in time fewer unknown or uncharacterised elements will be found in new versions and that this will allow greater control over the resulting performance, Cynthia retains a notable dislike for her wild relatives out there in nature. She, in contrast, has to remain in her place of birth, the laboratory, where she dreams of materialising her kind of virtual reality: one wherein her kind of control over life and nature takes the form of a world that is safe, secure, and sustainable, with little sense of her own complexity or of the environments that she might come to inhabit.

### *Cynthia's social life*

*Cynthia* does not mind having her behaviour checked by whomever has the technical expertise and motivation to get involved, whether inside or outside the laboratory. She expects to be watched over by others, which might even improve her designs by working bottom up and in ways that are transparent, co-operative, and decentralised. The problem is that it is mainly in a rhetorical sense that she gets to interact with the 'users' of her designs. This casts doubts on how 'open' she could become.

Consider the biosafety and bio-security of synthetic biology. These are the two themes that are most controversial, and an open community has been proposed as a more efficient means of dealing with the risks that accompany the ability to synthesise the DNA of polio, smallpox, extinct influenza strains, and dangerous pathogens (see ETC group 2007a, 2007b). In this model, preventing the risks linked to dangerous organisms that might fall into the wrong hands or that that be created and used with malicious intentions is seen as a responsibility of the scientists involved. Collaboration within the design community is proposed as more efficient for the detection of misuse than is regulation. This matches the well-known slogan of the Open Source movement: there is the need 'to create and maintain open networks of researchers at every level, thereby magnifying the number of eyes and ears keeping track of what is going on in the world' (Carlson 2003, 212).

The idea is that whenever there are enough contributors to a project, the problem must be easy to solve for someone. Following this viewpoint, resolving biosafety concerns depends on the availability of documentation and a transparent trial-and-error selection in the design process. Mistakes in the design of the BioBricks or other kinds of synthetic DNA being exchanged should be detected without the need to trust others not to make mistakes when using the technology (see Schmidt 2008). The kind of Open Source philosophy that is being invoked in synthetic biology as well as when separating the responsible kind of biohacking from possible misbehaviour is called Linus's law. This 'law' is named after Linus Torvalds, the inventor of Linux, and derived from a book called *The Cathedral and the Bazaar* that states that 'given enough eyeballs, all bugs are shallow' (Raymond 1999) where 'bugs' is a jargon term for errors and problems. The best known examples of how such bugs are solved through the diversity of the network is Linux, which was originally developed for personal computers. While only a small percentage of these run on Linux, it has by now become the leading operating system on mainframe computers and servers.

When comparing the field of synthetic biology to Linux, the former very comfortably invokes a notion of 'users' as innovators in contrast to manufacturers of various kinds, whether for software or for physical products (von Hippel 2005). The emphasis on a design community and a student competition closely resembles the Open Source philosophy that was coined in the nineties when the term 'open source' replaced the idea that free software would be available without any costs, which was considered inappropriate in the corporate world (see Berry 2008). Similarly, the BioBricks approach prioritises the technical achievements of the engineers who, like the programmer of source code, chooses 'open source' or 'closed source' depending on which of the two development models is more or less efficient under different circumstances (Berry 2004, 81; Raymond 1999).

Open source programming, however, is seldom easy, which is something that matters when taking the analogy with synthetic biology seriously. Linux is not necessarily that easy to use, and only recently are there versions that are no longer difficult to obtain, install, and operate for those without the necessary programming skills. It is not the simplest approach that should have priority: the analogy between synthetic biology and open source breaks down when considering its limits on the involvement for those users who are not experts to be the ones setting the priorities about what is a stable release and what is not. As is well known, the programming of Open Source Code, such as that of Linux, is much less susceptible to computer viruses than proprietary versions. For instance, the website of Ubuntu (a widely used implementation of Linux that is particularly user friendly) explains that it is necessary to be extremely careful in the design of a new version because 'regression in a stable release is a catastrophe'. This does not apply to synthetic biology, where the priority is placed on the detection of possible errors and misuse of its technology. There is no review process wherein priorities about reliability and stability are identified by collaborating closely with the users of its inventions. These might have very different ideas about the detection of errors, risks, and measures to prevent that the operating system might become unreliable.

Is it not the least to expect from an Open Source philosophy that it does not exaggerate its capacities and its reliability? Synthetic biology characterises the system or programmes that are studied primarily in terms of performance according to protocol. This reflects an exaggerated sense of control over the biological complexity of a synthetic construct that consists of mostly unknown and unstable biological processes. The analogy with Open Source that is being drawn in synthetic biology makes few efforts at truly opening up and revealing the messiness of biological processes, which disappears from view within its informatic rhetoric. The same applies to pressing questions of safety and security when these are turned into features of its design process. Hence, it is not the simplest approach that should have priority, nor should the involvement of experts who decide whether synthetic constructs are stable be confused with Open Source, especially in considering safety and security concerns or sustainability.

Finally, there is nothing in the BBF's setup that inhibits the corporate concentration of multinationals that has been characteristic of the plant biotechnology, chemistry, and pharmaceutical sectors. Even if the research tools are freely available and developed as Open Source projects, the results of the process are synthetic counterparts that can be patented separately and run on minimal cells that act as operating systems with biological functions that can plug and play food, energy, plastic, rubber, vanilla, and so forth.[6] What then of its 'open community' solving the world's problems by designing synthetic DNA? How open are you really when it is utterly unfeasible that the analogy with Linux comes to include the reliability of the 'releases'? Why not engage with the anxiety experienced by the potential users, whether related to biosafety issues or otherwise, rather than limiting direct involvement to potential synthetic biologists? This would require that the open community is truly inclusive and prioritises whomever has skills, tools, and knowledge that are important in a world wherein living materials are being translated into informatic formats.

## Conclusion

While Synthia and Cynthia sound the same, the difference in initial letter sets them apart as distinct personalities. While Synthia is the living embodiment of synthetic biology as a driver of the commodification of life, Cynthia's engineering vision of the biological future is anchored in Foucault's discussion of Cynic philosophy. Accordingly, the chapter began with a quote from the *Courage of Truth* wherein Foucault refers to Cynic practise as 'an extremely distinctive form of life'. Indeed, Cynthia is a truth teller, seeking to know the truth about life, which includes challenging whatever is not synthetic about herself and being principled in her demand for a life that is radically other. And while there are many types of cynicism and its history is far from uniform, simultaneously the cynic is 'the matrix, the embryo [. . .] of a fundamental ethical experience in the West' (Foucault 2011, 287).

To conclude, we first need to set her brand of cynicism apart from the kind of critique of the commodification of life that marks her sister's personality. This is not an easy task given that her preference for a more principled approach (to life) is taking its shape in a context in which the extension of engineering principles to biology and business principles to genome engineering is perceived as something inevitable. What difference can we really expect when many of those who could educate her about the world around her are heavily invested in the re-distribution of existing forms of political agency, subjectivity, and identity in support of the intensification of biological cycles of reproduction and functionality? The outcome is likely to be that her cynicism will take shape as a 'short cut to virtue' (ibid., 206), which is to say that she is difficult to believe when flouting her virtues.

On the one hand, these virtues are presented as simultaneously founded in nature and regulated through the truth seeking of the sciences. On the other hand, the short cut alternates between a rhetoric of responsibility and a cynic's preference for 'intolerable insolence', pointing out how the 'others are completely mistaken' in their understanding of the direct experience of nature (ibid., 313). For those who denounced her sister, it is probably inevitable for Cynthia's very existence to have meaning only in this regard: as that of the cynic who seeks to convince people 'to condemn, reject, despise, and insult the very manifestation of what they accept, or claim to accept at the level of principles' (ibid., 234). They would be correct in the sense that she has no patience for the conventional boundary between life and technology, whether as seeds, embryos, stem cells, or otherwise. However, this is not the only type of cynicism we have seen. We have to consider the radicalisation of synthetic life in its relation to 'the absurdity of the modern world' (ibid., 180).

It is the cynic who can show us that it is inevitable for us to find ourselves in proximity to synthetic biology and its understanding of man's relation to nature at the very moment when we might desire to distance ourselves from how new types of markets and ways to generate value emerge that reproduce vested interests. This is the reason for referring to Hugo Grotius, who, along with Hobbes and Rousseau, provides a point of reference to Foucault's problem of sovereignty (see Chapters 1 and 2). Grotius's 'fundamental re-examination of property in the colonial age' (Tuck 2003, 90), through its violence and race relations, shows the opening up of the natural world as the setting for such an 'essential' cynicism (Foucault 2011, 198).

What this means can be seen in the global scientific expeditions led by a heroic individual, the utopian communitarianism of the BioBricks Foundation, and Cynthia's existence. These are simply recent arrivals that begin passing on the long history of political theory, seamlessly fitting into an historical and geopolitical narrative about the natural world. Yet there is also the potential for becoming a different type of cynic: one who 'risks one's life, not just by telling the truth, and in order to tell it, but by the very way in which one lives' (ibid., 234). Our analysis shows that she is coming into existence at a time when very little remains inevitable. Established patterns are transforming rapidly, or, rather, their stability requires constant work on account of the immateriality of DNA and implies an intense contestation over the relationship to those who will have to live and work with her designs. This is where cynicism becomes critique. Her presence allows us to demonstrate the life sciences as a domain that is insulated from critique and to engage with the politics of science and technology from a vantage point at which we can demonstrate why it is that public discussion is unlikely to go beyond platitudes or is even impossible to conceive (Marcuse 1991, 11).

What is impossible to conceive, in the typical discussions of synthetic biology, is the re-materialisation of its information structure: how its re-organisation of how life is known and acted upon in synthetic biology brings us to a vision of the future wherein synthetic life runs on low-cost abundant biomass, which is like the cheap hardware on which one can indiscriminately run 'open or closed code'. Synthetic biology as a business seeks to profit from servicing informatic artefacts or from royalties and performance-based payments for the design of whatever biological commodity is in demand. From this perspective there is no need for an interest in the involvement of those who might become its users. Just as a Windows user does not necessarily know the operating system very well, the living materials on which the synthetic compounds are to run are standardised to be open to some rather than to others. Synthetic biology promises to run its genetic circuitry on a platform of standardised synthetic life forms that signals at a non-proprietary interface that prioritises the sharing of techniques and collaboration by those who opt into its way of doing things, primarily other synthetic biologists or experts in related fields.

Such a critical analysis makes it difficult to believe whether Cynthia's views are truly her own or instead that she is not really committed to openness and freedom in the sense in which Foucault referred to the 'kernel of Cynicism; practising the scandal of the truth in and through one's life' (Foucault 2011, 200). What is scandalous here is while the BBF foregrounds its community as open, it is limited in its ability and willingness to rethink how communities and user groups with different normative standards of truth and credibility are influenced by the transformation of the flow of knowledge, forms of exchange, and generation of value. While staying out of such networks means 'not to win' and to be excluded from the strategic resources mobilised around new commodity chains, Cynthia might end up regretting becoming known as another dystopian drama about how life forms (like herself) are being turned into commodities as means to manipulate bodies, minds, and the environment without any ends beyond itself.

Yet her relationship to her creators does not inevitably imply that Cynthia could not be convinced to change her mind no matter the circumstances. The question is whether she could be redirected to prefer a different approach: one that implies a re-orientation towards the need for a baseline of reciprocity between life scientists and those whose subjectivity is devalued when new types of exchange have become the standard. What began as a rhetorical strategy – the personification of a form of life in synthetic biology – is also a starting point from which to begin countering synthetic biology's self-image and how it advertises it. Cynthia is a means to identify that the potential for re-orientation, as was shown by engaging closely with different principles that seek a less possessive future.

The title of the chapter – synthetic life🄯 – is at once indicative of the need for proximity to synthetic biology, as a scientific practise and a business model, and of the need to be theoretically utopian with regard to the potential for otherness. We have taken only the first step, suggesting that we can exchange the modularity of Cynthia's designs and go for basic rather than easy, in much the same way that Linux's graphics do not run any unnecessarily complicated code that would restrict independent programming and that do not allow for control over the applications. Similarly, Cynthia should be much more honest and reveal how the genetically programmed behaviour of applications relies on unknown and unstable biological processes that might malfunction outside the laboratory in the face of its endless and messy biological processes, complex interactions, and reactions taking place in oceanic bodies, fields, and whatever other life forms she might encounter.

This is also the major hurdle that needs to be taken: Cynthia is not self-aware in dealing with the complexities that might undermine her confidence, nor is she in a place where she might acquire the skills she needs to go into the world, where real problems cannot be solved solely by sharing more data or creating access to techniques. We have seen her busy with convincing the public and the authorities to trust her. She has tried to appear virtuous by flouting that she can be technically fixed not to do anything against the law and staying close to everything new, modern, and dynamic about solving hunger, disease, energy shortages, and so forth. What is missing however, is a willingness to question how 'sharing' – as the norm for source code, seeds, plant biotechnologies, and whatever applications – can be extended to include and give voice to those who live and work with living materials and who wish to continue to do so on their own terms (see Deibel 2009, 2013; see also Chapters 4 and 7 of this book).

This is therefore who she is: she is someone who experiences the world as seen on a screen, where the images shown are under control and in which life is being simulated as immaterial until designs for safety, security, and sustainability become body again. The future belongs to her but only in the sense that she acts on a desire for an alternative route to become safer, more humanitarian, and environmental. Regrettably, such a path requires us to invent around, reverse engineer, and improve upon some of the really closed code – that of society – and release the potential for an 'open source society' (see Hardt and Negri 2004, 337–340). Perhaps we still have time to start becoming free, making new alliances needed for an alternative wherein the users of Open Source in the life sciences might be anyone with a relation to living materials, like those who grow it as crops, eat it as food, take it as medicine, or – by virtue of their bodies – are it.

If she's got it – which she might – she could, perhaps, come to realise that there are different lives being led on the outside of her confinement of her home in the laboratory. With guidance she might become a safer human-loving and Open Source kind of cynic whose nature manifests itself as an impulse towards philosophy that runs 'through the whole of western history' (Foucault 2011, 174). Despite these odds, and as far as her future is concerned, we may yet establish a radically different kind of interface: one aimed at bringing out the lives of others as integral to the life forms that inhabit and are produced in laboratories and digital environments. And for that – dear synthetic biologists, dear ethical biohackers – what would be really useful is to begin reverse engineering 'Life™' into lives that are free.

## Notes

1 For the Press Release, see www.jcvi.org/cms/research/projects/synthetic-bacterial-genome/press-release/ (accessed on 1.3.2016).
2 For information on the Blue Brain Project, see http://bluebrain.epfl.ch/ (accessed on 1.6.2016).
3 For details of the NeuroCommons project, see http://neurocommons.org/page/Main_Page, and for some information on the HapMap project, see http://hapmap.ncbi.nlm.nih.gov/thehapmap.html (both accessed on 1.5.2016).
4 See https://biobricks.org/bpa (accessed on 1.3.2016).
5 For some impression of these or similar examples, see http://igem.org/About (accessed on 1.3.2016).
6 For further discussion and analysis, see the overview presented on the ETC Group Web site, at www.etcgroup.org/issues/synthetic-biology (accessed on 1.5.2016).

## References

Arneil, B. (1996) *John Locke and America: The Defence of English Colonialism*. Oxford: Clarendon Press.

Baker, D., Church, G., et al. (2006) Engineering Life: Building a Fab for Biology. *Scientific American*, 294: 44–51.

Benkler, Y. (2006) *The Wealth of Networks*. Available at: www.benkler.org/ (accessed on 1.12.2016).

Berry, D. (2004) The Contestation of Code: A Preliminary Investigation Into the Discourse of the Free/Libre and Open Source Movements. *Critical Discourse Studies*, 1(1): 65–89.

Berry, D. (2008) *Copy, Rip, Burn: The Politics of Copyleft and Open Source*. London: Pluto Press.

Brent, R. (2004) A Partnership Between Biology and Engineering. *Nature Biotechnology*, 22(10): 1211–1214.

Bügl, H. and Danner, J.P., et al. (2007) DNA Synthesis and Biological Security. *Nature Biotechnology*, 25(6): 627–629.

Calvert, J. and Joly, P. (2011) How Did the Gene Become a Chemical Compound? The Ontology of the Gene and the Patenting of DNA. *Social Science Information*, 50(2): 1–21.

Canton, B., Labno, A., and Endy, D. (2008) Refinement and Standardization of Synthetic Biological Parts and Devices. *Nature Biotechnology*, 26(6): 787–793.

Carlson, R. (2003) The Pace and Proliferation of Biological Technologies. *Biosecurity and Bioterrorism: Biodefence Strategy, Practice and Science*, 1(3): 203–214.

Carolan, M.S. (2010) Mutability of Biotechnology Patents: From Unwieldy Products of Nature to Independent Object/s. *Theory, Culture & Society*, 27(1): 110–129.

Caulfield, T. (2011) Reflections on the Gene Patent War: The Myriad Battle, Sputnik and Beyond. *Clinical Chemistry*, 57(7): 977–979.

Church, G. and Regis, E. (2012) *Regenesis: How Synthetic Biology Will Reinvent Nature and Ourselves*. New York, NY: Basic Books.

Deibel, E. (2009) *Common Genomes: On Open Source in Biology and Critical Theory Beyond the Patent*. PhD dissertation. Available at: http://dare.ubvu.vu.nl/handle/1871/15441 (accessed on 1.12.2016).

Deibel, E. (2013) Open Variety Rights. *Journal of Agrarian Change*, 13(2): 282–309.

Delfanti, A., Castelfranchi, Y., and Pitrelli, N. (2009) What Dr. Venter Did on His Holidays: Exploration, Hacking, Entrepreneurship in the Narratives of the Sorcerer II Expedition. *New Genetics and Society*, 28: 415–430.

Dutfield, G. (2003) *Intellectual Property Rights & the Life Science Industries: A Twentieth Century History*. Aldershot: Ashgate.

ETC Group. (2007a) *Patenting Pandora's Bug: Goodbye, Dolly . . . Hello, Synthia!*. Available at: www.etcgroup.org/content/patenting-pandora%E2%80%99s-bug-goodbye-dollyhello-synthia (accessed on 1.12.2016).

ETC Group. (2007b) *Extreme Genetic Engineering: An Introduction Into Synthetic Biology*. Available at: www.etcgroup.org/content/extreme-genetic-engineering-introduction-synthetic-biology (accessed on 1.12.2016).

Forster, A.C. and Church, G.M. (2006a) Synthetic Biology Projects in Vitro: Synthesizing Self-replication and Life. *Genome Research*, 17: 1–6.

Forster, A.C. and Church, G.M. (2006b) Towards Synthesis of a Minimal Cell. *Molecular Systems Biology*, 2(45): 1–10.

Foucault, M. (2011) *The Courage of Truth*. New York, NY: Palgrave Macmillan.

Franklin, S. (2007) *Dolly Mixtures: The Remaking of Genealogy*. Durham, NC: Duke University Press.

Haraway, D.J. (1997) *Modest_Witness@Second_Millennium.FemaleMan©_Meets_Onco mouseTM: Feminism and Technoscience*. New York, NY: Routledge.

Hardt, M. and Negri, A. (2004) *Multitude: War and Democracy in the Age of Empire*. New York, NY: Penguin Press.

Harvey, D. (1996) *Justice, Nature & the Geography of Difference*. Oxford: Blackwell Publishers.

Hayden, C. (2003a) From Markets to Market: Bioprospecting's Idiom of Inclusion. *American Ethnologist*, 30(3): 1–13.
Hayden, C. (2003b) *When Nature Goes Public: The Making and Unmaking of Bio-Prospecting in Mexico*. Princeton, NJ: Princeton University Press.
Helmreich, S. (2008) Species of Biocapital. *Science as Culture*, 17(4): 463–478.
Helmreich, S. (2009) *Alien Ocean: Anthropological Voyages in Microbial Seas*. Berkeley, CA: University of California Press.
Henkel, J. and Maurer, S.M. (2007) The Economics of Synthetic Biology. *Molecular Systems Biology*, 3(117): 1–4.
Hilgartner, S. (2012) Novel Constitutions? New Regimes of Openness in Synthetic Biology. *BioSocieties*, 7: 188–207.
Hoare, A.L. and Tarasofsky, R.G. (2007) Asking and Telling: Can 'Disclosure of Origin' Requirements in Patent Applications Make a Difference? *The Journal of World Intellectual Property*, 10(2): 149–169.
Kelty, C.M. (2004) Culture's Open Sources: Software, Copyright, and Cultural Critique. *Anthropological Quarterly*, 77(3): 547–558.
Kelty, C.M. (2012) This Is Not an Article: Model Organism Newsletters and the Question of 'Open Science'. *BioSocieties*, 7(2): 140–168.
Knight, T.F. (2005) Engineering Novel Life. *Molecular Systems Biology*, 1(1): 1.
Mackenzie, A. (2010) Design in Synthetic Biology. *BioSocieties*, 5: 180–198.
Mackenzie, A., Waterton, C., et al. (2013) Classifying, Constructing, and Identifying Life: Standards as Transformations of the Biological. *Science, Technology and Human Values*, 38(5): 701–722.
Marcuse, H. (1991) *One-Dimensional Man: Studies in the Ideology of Advanced Industrial Society*. London: Routledge.
Newman, S.A. (2012) Synthetic Biology: Life as App Store. *Capitalism Nature Socialism*, 23(1): 6–18.
Parry, B. (2004) *Trading the Genome: Investigating the Commodification of Bio-information*. New York, NY: Colombia University Press.
Pottage, A. (2006) Too Much Ownership: Bioprospecting in the Age of Synthetic Biology. *BioSocieties*, 1(1): 137–158.
Rai, A., and Boyle, J. (2007) Synthetic Biology: Caught Between Property Rights, the Public Domain, and the Common. *PLoS Biology*, 5(3): 389–393.
Rathenau Institute. (2006) *Constructing Life: Early Social Reflecting on the Emerging Field of Synthetic Biology*. The Hague: Rathenau Institute.
Raymond, E.S. (1999) *The Cathedral and the Bazaar: Musings on Linux and Open Source by an Accidental Revolutionary*. Cambridge, MA: O'Reilly.
Rimmer, M. (2009) The Sorcerer II Expedition: Intellectual Property and Biodiscovery. *Macquarie Journal of International and Comparative Environmental Law*, 6: 147–187.
Rousseau, J.J. (1993) *The Social Contract and Discourses*. London: Everyman.
Schmidt, M. (2008) Diffusion of Synthetic Biology: A Challenge to Biosafety. *Systems and Synthetic Biology*, 2(1–2): 1–6.
Tamminen, S. and Brown, N. (2011) Nativitas: Capitalising Genetic Nationhood. *New Genetics and Society*, 30: 73–99.
Tuck, R. (2003) *The Rights of War and Peace*. Oxford: Oxford University Press.
van Dooren, T. (2007) Terminated Seed: Death, Proprietary Kinship and the Production of (Bio)Wealth. *Science as Culture*, 16(1): 71–94.
Verne, J. (1992) *Twenty Thousand Leagues Under the Sea*. Hertfordshire: Wordsworth Editions.
von Hippel, E. (ed.) (2005) *Democratizing Innovation*. Cambridge, MA: MIT Press.

# 7 Re-thinking the age of biology

## Biomass, biohacking, and open-source seeds

> What I have been trying to show over the last few years is certainly not how, as the front of exact sciences advances, the uncertain, difficult and confused domain of human behaviour is gradually annexed by science [. . .]. We should be looking for a new right that is both antidisciplinary and emancipated from the principle of sovereignty.
>
> Foucault, *Society Must Be Defended* (2003, 39–40)

### Introduction

The basic formula of the global bio-economy promises that a renewal of industrial society could become reality in the near future as a result of the new ways to make use of biomass as a basic material that replaces fossil fuels. Different terms are being used, sometimes it is called a 'knowledge-based bio-economy' (the European Commission), a 'comprehensive bio-economy strategy' (Germany), 'a biobased economy' (the Netherlands), a 'bio-economy blueprint' (the US), and biomass industrialization (Japan). Instead of 'biomass', a specific type of biological resources might be mentioned, or only certain sectors might be discussed. Regardless, technological change is presented as integral to a transition that affects all of us, includes all of us, and leads us to a sustainable society.

Particularly in the EU and its member states, there are numerous policy processes attempting to bring together stakeholders and engaged participants in a continuous interdisciplinary dialogue among industry, science (both natural sciences and humanities) and civil society. Various instruments are being explored, like sustainability criteria, certification, licencing, 'green deals', innovation contracts, partnerships, and the like. Invariably, the policy language invokes a demos, deriving strength from its ability to accommodate diverging and converging interests, diverse points of view, and multiple understandings of the transition to sustainability (Frow et al. 2009). Such inclusiveness, however, implies a politics of imagination that itself requires examination in so far as it re-establishes the principle of sovereignty for the age of biology that is being promised. Biology and sovereignty come together, as Foucault explains, by how the power of disciplinary knowledge that is exercised through the social body is shaped as a 'grid' that 'cannot in any way be transcribed in right, even though the two necessarily go together' (Foucault 2003, 37).

The first section that follows examines the coherence of the global bio-economy, as a concept, against the backdrop of the life sciences as a field that is built on fragmented facts, data, and artefacts. What this implies is that it is a resolutely political vision that shows a socio-political order that is being imagined as a unity of concrete, discrete, and consolidated innovations and policies. The second section echoes the quotation's 'new right' by discussing 'biohacking' and 'open-source seeds'. Each of these can be considered as counterexamples and challengers to the suggestion of a global bio-economy as an alternative. More precisely, the two topics are both related to the language of openness and access but are different in how they seek to re-configure and re-distribute forms of agency, subjectivity, and identity. Examining these three attempts at formulating alternatives in various ways will allow us to return to Foucault's argument that 'we cannot go on working like this forever; having recourse to sovereignty against discipline (ibid., 39). The aim of this chapter is therefore not just a critical examination of these fields, re-establishing a boundary between the 'reality of things' and 'idealism'; rather the notion of the alternative is key to understanding how the boundary between life and society is being re-imagined constantly. As theorised earlier in this book: what counts as an alternative should not be secondary to critique but an integral part of the analysis.

## The global bio-economy as an origin myth

### *Species of biocapital*

Any 'new right' or alternative politics of imagination has to find a way to combine a critique of biocapitalism with a 'refusal to make capital into the coin of exchange' (Helmreich 2008, 475).

What this 'coin' refers to is that the result of prioritising the political economics of the life science is usually a critique of (bio-)capital wherein alternatives are barely imaginable, remote, idealistic, and at a distance from the core dynamic of the political economy of the life sciences.

This so-called 'biocapital school' is a reference to an article by Stephan Helmreich called 'Species of Biocapital' (2008), which identified a body of literature that examined the life sciences by combining political economy and STS. He explains that:

> [t]he term, paging back to Marx, fixes attention on the dynamics of labour and commodification that characterize the making and marketing of such entities as industrial and pharmaceutical bioproducts. [. . .] Biocapital also extends Foucault's concept of biopolitics, that practice of governance that brought 'life' and its mechanisms into the realm of explicit calculations.
>
> (Helmreich 2008, 464)

He discusses a spectrum of texts published as recently as the mid-2000s that appeared in relatively rapid succession. These include Sunder Rajan's *Biocapital* (2006), Franklin's *Dolly Mixtures* (2007), Cooper's *Life as Surplus* (2008), and others.

Seen together, the approach of the school is controversial. Critics correctly object to the suggestion that there is anything special about the outer reaches of knowledge production in the life sciences and the opening up of biological terrains to market forces. Notably Birch and Tyfield's article 'Theorizing the Bio-economy, Biovalue, Biocapital, Bioeconomics or . . . What?' (2013) accuses the approach of fetishising everything biological. The authors argue that there is nothing specific to the life sciences when examining its relation to neoliberal capitalism and the restructuring of the capturing of profits in line with the world of high finance. Such a critique is valid, of course, but only to the extent that it challenges the hypothesis that the making and marketing of bio-objects in various fields is a principle influence on how established patterns of finance, trade, and industrial production unfold everywhere else. Such a critique simply mirrors the more general point about studies of the politics of science and technology, the argument that STS studies are too strictly tied to analysis of performativity in isolated case studies, mostly in the context of the laboratory.

The problem with such a critical position is that the priority is unreservedly given to the economic and geopolitical arrangements of neoliberalism. From a methodological standpoint the study of the biopolitical is highly relevant because it is able to foreground that the global bio-economy is not an achieved state of affairs but refers to multiple, competing forms of life that co-exist within policy framework with politically and legally unstable and often unpredictable outcomes. While critics of the biocapital school are right in calling for deeper engagement with contemporary scholarship on value in capitalism in the analysis of such multiplicity (e.g., Birch and Tyfield 2013; see also Lazzarato 2014), this does not lead away from Foucauldian scholarship but towards opening up of the theoretical premises underlying the literature examining biocapitalism.

This has been one of the objectives of this book, and also more generally, we would agree that what is pressing is a renewal of sociological approaches able to go beyond analysis of collections of individual cases. Earlier in this book we sought to contribute by engaging with the global politics of genetic resources (in Chapters 4 and 5). This set the backdrop for the examination of experimentation in synthetic biology, challenging the same divide between critique and multiplicity. In this chapter a subsequent step can be taken: moving from a critique of the 'global bio-economy' to rethinking what counts as an alternative in contemporary biocapitalism. In this regard it is particularly important to notice the underlying synthesis of Marx's work on value with Foucault's symmetry of life, labour, and language. The implication of the latter is the suspension of any *a priori* preference for one field over any of the others. Doing so is crucial when seeking to place the potential for alternatives at the centre of a critique.

Exemplary in this regard is Cooper's *Life as Surplus* (2008). The contemporary development of the life sciences is situated in the context of how the US responded to the crisis of its industrial model by abandoning the gold standard and has since then become the world's largest debtor. Accordingly, one of the primary ways wherein this debt was used was the financing of the life sciences. Such emphasis on debt creation also implies that the life sciences are at the core of the ongoing renewal of capitalism. The life sciences are developed as a promise on the repayment of state debts and are therefore implicated in the temporality that neoliberalism enforces on the present. Specifically, Cooper argues that 'profits will depend on the accumulation of biological futures' (ibid., 24). This refers to the many kinds of business models in the life sciences wherein 'biological, economic and ecological futures' are 'intimately entwined' as subjects of speculation (ibid., 20).

As impressive as the analysis has been, it is nonetheless a startling moment when coming to the realisation upon turning the last page of her final chapter that there is nothing but critique. It suddenly ends with a brief commentary to highlight that the biological future might not belong to the US, that it might fail to capture the profits that were expected to resolve the debt crisis of the US industrial model. Especially noteworthy is the absence of any kind of discussion of the consequences of the emergent model that she has described, which appear to be nothing short of a disaster whether or not the US manages to capture any profits. The brevity of her conclusions gives reason to reconsider her analysis along with the biocapital literature as a whole.

We agree with her that a Foucauldian analysis implies symmetry among life, language, and labour. This is one reason for the abrupt conclusions. Neither privileging biology over economics and language or the reverse means that it is not possible to end up immobilising developments in the life sciences by privileging a Marxist critique of political economics. After all, Foucault argued that critiques (such as that of Kant and Marx) introduced no real discontinuity within epistemological arrangements of Western knowledge of the nineteenth century. Instead critique was a part of it, relied on it, and had no power to exercise over it. Like Foucault's 'The Order of Things', Cooper offered a critique of critique, a position that is valid in its insistence on the indeterminacy of the future that is being enabled in the life sciences. The problem, as stated before, is that such a position implies that alternatives become idealistic and distant rather than closely tied to a politics of imagination.

Consider once more the rhetoric surrounding the global bio-economy. This concept seeks to circumvent how it has become exceedingly complex to determine who decides about the usage of crops, biological resources, or related inventions. The smooth transition to a sustainable society that is being imagined is political in how it legitimises, regulates, limits, and dismisses states of living in, being in, and acting upon imagined worlds that could become real. What kind of critique should be the response to such an affirmation of the need for an alternative? Can we simply affirm the indeterminacy of the linear and decisive ways wherein the biological future is imagined as refashioning the world? In our view, it is not sufficient to argue in favour of complexity and multiplicity. Its implication is that there is no contest over the politics of imagination on its own grounds: critique leaves the politics of imagination to belong exclusively to the forces of neoliberal capitalism. While we acknowledge the need to engage directly with the global bio-economy's close ties to established patterns of capitalism, we prioritise a critical examination of the concept as a redemption story that promises a return to nature. It is in these mythical terms that its power effect is exercised: the rhetorical isolation of the continued convergence of synthetic biology and chemistry from the irresponsibility of petrochemical industries. Re-establishing this continuity is not solely about political economy and industries that are merging; it allows us to show how neoliberalism is being re-imagined in synthetic biology and by extension in many of the related issue areas.

Ultimately, the alternative is not just performed in modern biology but in many different sites, each of which is characterised by a relation to the forces of neoliberal capitalism. It is by thinking across these relationships (in space and time) that such alternatives are no longer a secondary part of a critical analysis but constitutive of the terrain of contestation where multiple, competing, forms of life, knowledge and related modes of governance co-exist, compete, and materialise exchange and value from socially heterogeneous practises.

### *The biological age as a return to nature*

With some regularity the global bio-economy is presented as a return to its natural foundations. This is reflected in the image in Figure 7.1, used for the public relations of DSM,[1] a Dutch chemical-industry and multinational (once Dutch State Mines).

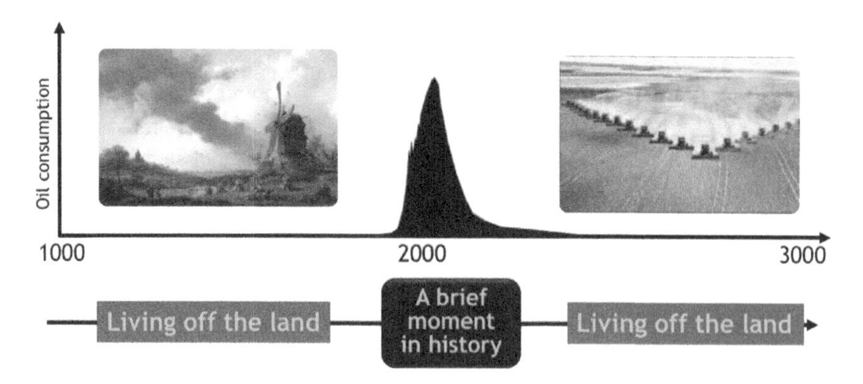

*Figure 7.1* From the fossil age to the bio-age (DSM)

This image was originally created by DSM chemicals and used in various public relations activities.[2] Permission was granted on 18 January 2017, personal e-mail communication.

What the image below shows is the biological age as the end to 'a brief fossil moment in history'. This 'end' is represented with a Dutch windmill in a pastoral setting, which subsequently takes on a scale that matches the massive cultivation of corn as a reflection of DSM's involvement in the refining of ethanol in the US (see also Veraart et al. 2011). The message is that the unity of time requires a biological age, as a return to normality and nature in the shape of a world wherein biological resources once again are the basis for food, heat, transportation, construction, and so on. Re-establishing the unity of time requires the realisation of a long wish list of innovations: generating warmth and electricity from waste, driving cars on fuels that are based on sugar or algae, while bottles, bags, and every other type of plastic are made from renewable and degradable materials.

More often, however, the slogans are naturalistic rather than historical, with the slogans of companies offering chemicals that are 'naturally designed and engineered' (GreenBiologics), 'powered by nature' (Corbion), as an entry into 'a new and greener world of infinite possibilities' (BioAmber), and so on.[3] Such naturalistic language is fairly well established, although others prefer to simply highlight their technical expertise (Gevo, Intrexon, Avantium).[4] What the naturalistic rhetoric reflects is how the global bio-economy combines growth-focused developmentalism with limits-focused environmentalism. Each of these mobilises a biological metaphor. On the one hand, the term 'development' came to refer to the social sphere only at the end of the eighteenth century, interweaving the Hegelian concept of history with the Darwinist concept of evolution. To develop is then a natural process with limits, as form fits size and size fits form, and failure to develop is an anomaly (underdevelopment). The language of ecological limits, on the other hand, implies a contrasting view, wherein growth has an end-point, just as with the human body; when a certain size is surpassed further economic growth might have catastrophic effects. A combination of the two implies mixing the metaphors of infinite growth with a contrasting view wherein production and reproduction are natural processes that after a certain point turn growth into an abnormality (like with cancer or obesity; see Mert 2012, 182–183).

In the policy languages of the global bio-economy this tension takes a variety of forms, but the rhetoric centres on the notion that there should be 'intelligent' use of biomass. Unsustainable situations are identified primarily with society's dependence on the fossil-fuel industry, which leads to environmental degradation, resource depletion, and conflict. Similarly limits are reached when biomass is used unintelligently, resulting in famine, malnutrition or impoverishment or exacerbating climate change. Intelligence, on the other hand, requires more coordination of policies across regions, sectors, and countries as well as political decision making that guarantees that the transition to sustainability becomes an irreversible process wherein it is impossible to regress to today's state of affairs.

As Roland Barthes (1987 [1972], 117) reflects, myths are metaphors that 'transform history into nature'. In this register the merger of growth and limits is a myth of origin that can be seen as the foundation for the implementation of new social agreements and standards of credibility that cut across natural and social worlds (see Serres 1995; see Latour 1993). There are many such 'Green New Deals', particularly in Europe, one of which is the 'Manifest for a Biobased Economy', which was signed in the Netherlands in 2011 by forty-three stakeholders from industry, science, policy, and civil society organisations. All agreed that it offered opportunities to combine prosperity with a reduction of greenhouse gas emissions and the ecological footprint and to create greater access to sustainable means of living. Success is invariably seen as conditional on inclusive and responsible innovation models, proper communication, and the collaboration of various state and non-state stakeholders.[5] The document mentions the 'cascade model', which argues that biomass should not be used where there are other sustainable solutions available based on solar, wind, or water technologies. Only those applications that combine high value with high ecological benefits should be prioritised, which requires closing the cycles of resources being used (their cultivation, processing, consumption, recycling, etc.).

The suggestion is that there is a 'contemporary bioconstitutional moment', an attempt to 'restore epistemic and normative order under uncertain circumstances' (Jasanoff 2011, 3–4). Accordingly, the return to natural foundations functions as a redemption story wherein industrialised societies are saved from crisis and decay takes shape as a social contract that requires the continuous balancing of size and form in a natural and continuous process. It reaches conceptual optimum or equilibrium as a perfectly coordinated circular economy of renewable resources cascade models and stakeholder inclusion.

However, the mythical frame presumes a citizenry. The citizens of this combination of social and natural harmony are imagined as the inhabitants of sustainable societies wherein an affluent population live their lives in perfect harmony with each other and nature, who out of their own free will and desire for goodness choose to reciprocate how they are being imagined as working in the new centres of production, living around the corner, drinking from a (bio-)plastic bottle, or taking new foodstuffs out of bioplastic bags.

### *The life sciences and the 'new' chemistry*

The transition to sustainability is imagined, in its first instance, as an end to the age of oil. After all, this alternative is not only a promise of a better future but a reaction to a negative state of being. The state of society is one of dependence on the usage of fossil fuels and industries that continue their refinement and extraction from mines and oil wells. This includes an intense search for new deposits that can only now be accessed using new techniques. Among the areas being opened up are deep-water drilling in the oceans and oil and tar sand in the Arctic as well as heavily populated areas with shale gas underneath. When presented as such a stark choice the direction might appear obvious, but what conveniently disappears from such a politics of imagination is that a similarly stark choice might be found within the life sciences. In other words, the type of issues that seem urgent changes drastically when we ask not only whether society should escape from its dependencies but question the direction we should escape to and how we should go about organising it.

Consider LS9, which was until recently a synthetic biology start-up but was then sold for half the original investment. The result has been that its technologies are no longer used in line with its slogan: *the best replacement of oil is oil*. What LS9 did was produce renewable diesel by using microorganisms. One of its founders, George Church, one of the most famous synthetic biologists, explained the benefits over conventional biofuels such as ethanol by pointing out that the latter will have to compete with food crops. Ultimately you need to 'burn more of it to get the same amount of energy' (Church et al. 2012, chapter 4). The refinery process requires energy, which defeats the purpose of 'growing petroleum', which is to leave oil and gas in the ground and 'clean the atmosphere'. He prefers using microorganisms like cyanobacteria, algae, and *E. coli* to create synthetic diesel that is chemically closer to conventional diesel, gasoline, and jet fuel (ibid.).

LS9 was attempting to create a fuel that could be used more widely than biofuels that are blended with conventional fuels or that require special infrastructure. What went wrong? The company wanted to develop an industrial biotechnology platform by scaling up an engineered metabolic pathway (that utilises *E. coli* to manufacture fatty acids). This required the ability to predict and control how the microbes behave in industrial environments rather than in laboratories. In addition, the process depended on sugar as a raw material. Oil prices were low at the time, as were the profit margins in the petrochemical industry, so the price of sugar was prohibitive, and large commercial plants were necessary for the production of volumes large enough to be competitive. Societies are still highly dependent on fossil fuels, but more importantly, this dependence will cross-cut the realisation of the vision of the biological future. What the case of LS9 shows is how closely linked the global bio-economy is likely to be to the petrochemical industries that it seeks to replace and how unlikely a clean 'break' is. The renewal of industrialised society, its return to nature, will take shape within the narrow confines of competitiveness of global commodity markets dominated by mass-produced, cheap products wherein territories and labour are interchangeable. Accordingly the industrialisation of synthetic biology is only possible for certain activities and areas of interests; the competition with existing sectors implies that other values are side-lined. Whatever is not valued by global markets will not seem to be competitive, like water quality, meaningful employment, or the landscape (Brunori 2013; Levidow et al. 2013).

Obviously this is not a new problem. The business model of synthetic biology and its industrialisation is a successor to the incorporation of the life sciences into the patenting regime that had emerged in chemistry. As also discussed in Chapter 6, patenting had developed in response to advances and industries related to organic chemistry in the late nineteenth century, including the appearance of research laboratories in corporations (Dutfield 2003). Accordingly the globalisation of DNA patenting in the 1980s and 1990s extended the technical criteria for the isolation and purification of chemical compounds. A bacterium was legally a chemical, a composition of matter, while before it had not been patentable, as it existed already and had been considered a representation of nature (Calvert and Joly 2011; Carolan 2010; see also Chapter 4). DNA patents were integral to the Trade-Related aspects to Intellectual Property Rights (TRIPs) agreement, part of the Uruguay Round on the General Agreement on Tariffs and Trade (GATT) concluded in 1993. This was unprecedented, because the newly established World Trade Organisation (WTO) would be responsible for the administration of patents: the WTO has the authority to investigate compliance with its rules, principles, and minimal standards, and this authority is backed up with a trade-dispute settlement mechanism (May 2000; Sell 2007).

Furthermore the liberalisation of agriculture at the global level occurred at the same time. Up until then, GATT Agriculture had not been governed as a trade mandate, but primarily by the FAO. Henceforth it was also considered to be within the jurisdiction of the WTO, which aims to facilitate the harmonisation of markets for agricultural produce as well as the goods and services that are protected with intellectual properties. It is this combination of globalising patents and agriculture in the 1980s that made possible a business model that combined the sale of chemicals used to cultivate crops with exclusively owned seeds. As was well documented at the time, there has been an unprecedented corporate concentration and integration of commodity chains for crops, plant biotechnologies, and chemicals (Mooney 1979; Yoxen 1984; Goodman et al. 1987; Pistorius and van Wijk 1999). Synthetic biology is following a similar pattern, opening up new biological terrains to investments from chemical companies that are again trying to renew their business models so as to retain a competitive hold over global value chains.

Some two decades after the previous round of chemical companies' convergence with the life sciences, only a few genetically engineered traits are in widespread use, and these have as their primary feature that they make crops more resistant to chemicals (e.g., herbicides). Simultaneously, agriculture and patents are still among the principle areas of dispute of the current trade negotiations in the Doha Round (Lee and Wilkinson 2007). It is unlikely that the Doha Round will conclude any time soon; certain countries are not giving up subsidies for agriculture, and the rest of the world is in no hurry to protect the intellectual properties on goods and services that are considered vital to the US and EU economies. Whatever the outcome might be (whether strengthening intellectual properties, liberalising agriculture, or both), it will remain closely tied to the commodification of plant varieties. There is no outcome on the table that does not revolve around monocrop varieties, either in industrialised agriculture using conventional crops or as genetically modified cash crops that can be traded globally.

What the sell-off of LS9 reveals is the extent to which this situation is characteristic of synthetic biology and its attempt at ecological modernisation. On the one hand, the industrialisation of synthetic biology seems to mirror the promise of plant biotechnology, delivering 'miracle crops' that would bring more food, more fuel, and whatever else. On the other hand, synthetic biology is oriented towards a global bio-economy as if biological materials can be seen as a basic resource that is detached from the materiality of species of plants and the embodied-know-how about living and working within multi-species environments. For example, sugar, which is a global agricultural commodity, was integral to the suggestion that there can be 'renewable diesel' even though such 'biomass' seeks to replace conventional fossil fuels with a further intensification of industrialised agriculture.

LS9 might have folded, but a similar business model might be more successful, industrialising the production of fuel without too much agricultural input or other types of competition over resources (like land and water). Also in such a scenario, however, the convergence with chemistry implies a brand of bio-economics that proposes a comprehensive alternative that consigns to its margins any visions of sustainability that are not (or cannot be) valued by global commodity markets. Sometimes the model relies on microorganisms, and other times biomass is cheap because it is based on conventional cash crops. In either scenario the viability of culturally diverse ways of life that rely on and sustain the diversity of plant life cannot be articulated as a key priority within the context of the policy language on the global bio-economy. Lots of issues might be mentioned, but as secondary concerns within the confines of the underlying vision for a sustainable future. Ironically we are again supposed to 'live off the land', as DSM chemicals phrased its myth of origin, yet such a return to nature in none of its features suggests any type of practical commitment to shortening supply chains, re-territorializing biological resources or other types of measures that show that culturally and biologically diverse ways of life are a precondition for sustainable societies.

## The (brave new) world of the new biology

### *Escaping the Savage Reservation*

> they believed in their hearts that the solution to every problem – whether psychological, sociological or more broadly human – could only be a technical solution.
>
> Michel Houellebecq, *Elementary Particles* (2000, 262)

With some regularity the protagonists of the field of synthetic biology present themselves as the pioneers of a new industrial revolution and advocate an engineering philosophy that has as its ideal to perfect and control cellular reproduction. For example, prominent synthetic biologist George Church (who wrote early on about the minimal cell project discussed in Chapter 6) claims a new age of biology is upon us. Here the language of industrial revolution replaces that of the break with the age of fossil fuels, but the result is similar: a new age of biology. Furthermore he predicts that synthetic biology implies 'the power to control our future biological development – to understand and then manipulate the evolving genome of life itself' (Church et al. 2012, chapter 7). To think through the implications of such a transhistorical sense of control, articulated by one of the most widely known synthetic biologists, it is useful to turn to speculative fiction, which has the ability to affirm such a shape of life to come. *Elementary Particles*, the book cited above, is particularly brilliant in how it dramatises the moment when mankind decided it 'should control the evolution of the world as a whole – and in particular its own biological evolution' (Houellebecq 2000, 332).

On its final pages, the reader is told that the purpose of the novel was to honour the courage of the first species in history capable of creating the conditions for its own replacement. A witness to the changes taking effect comments at the end of the novel that 'it is even surprising to note the meekness, resignation, perhaps even secret relief with which humans have consented to their own passing' (Houellebecq 2000, 263). The novel salutes the tormented, egoistic, cruel, and violent species, barely distinguishable from monkeys, that in its unhappiness was brave enough to bring an end to the gradual dissolution and fragmentation of reason in the materialistic age (ibid., 263–264). The narrative of the novel hinges on a moment of breakthrough in the life sciences, the definitive and final advance, a paradigm change in biology: perfect replication of genetic code, no matter how complex, which implies that 'every cell contained within it the possibility of being infinitely copied. Every animal species, however highly evolved, could be transformed into a similar species reproduced by cloning, and immortal' (ibid., 258).

Such a decisive paradigm change is unrealistic, the novel presents an exaggerated and overtly singular account of the biological future. Yet the same rhetoric surrounds synthetic biology as seen, for instance, in George Church's and Craig Venter's idealisation of the comparisons between biological reproduction with computing, engineering, and industrial revolutions (industrial, computation, nano, synbio, etc.). For Houellebecq it is, of course, a literary device; it enables him to dramatise the perpetually distant horizon of scientific progress and technology, turning it into a moment when the continued existence of humanity has become a clear choice. The novel describes the period leading up to the exact moment when mankind becomes able to sever itself from genetic individuality as the principal cause of human misery, the loss of youth, disease, old age and death.

Houellebecq ruthlessly exposes the emptiness of the contemporary popular culture when he prioritises the tragic longing to escape the crisis of the present. It is humanity in crisis that explains the appeal of technological utopias, particularly the imagination of a future wherein man is liberated from his own biology. Biology is part of a vast historical stream that draws humanity and every living being in its current, arriving at a destination that is exactly where the reader might have suspected that we were heading: the destination was always Aldous Huxley's Brave New World. The novel explains why the Brave New World came into existence and one of the main characters tells us what is happening when he observes that

> Huxley's world is usually described as a totalitarian nightmare, an attempt to pass the book as a vicious accusation. That's most hypocritical. On all points, genetic control, sexual freedom, the struggle against old age, its culture of leisure – Brave New World is a paradise to us, in essence it is exactly the world we are trying to achieve, until now without success.
>
> (Houellebecq 2000, 123)

Unlike Huxley's great novel, Houellebecq does not accuse the sciences for how they undermine the autonomous subject, equipped with rights, reason, and dignity. Instead he shows how compelling it is to escape from today's individualism and misery, placing us in the position of John Savage, the main character of Huxley's *Brave New World*, at exactly the moment he must choose whether or not to leave the 'Savage Reservation'. The surname and the location are not incidental – they echo the savage of early-modern theories on the natural foundations of society. John Savage is a tragic hero that greatly resembles Rousseau's savage. Also the latter has to leave the state of nature once coming into contact with civilisation. John Savage too leaves the state of nature to enter into a state of society and does not really have a choice about it; just like Rousseau's savage, he has no choice but to leave.

The difference, of course, is that John Savage does not embrace the technoscientific civilisation while Man in *Elementary Particles* embraces the biological future. Rather than approaching the transition with enthusiasm, as is the case with post- or trans-humanisms, Houellebecq's escape towards scientific paradise is solely that: a withdrawal from this reality. John Savage chooses to die a futile death over joining with the Brave New World, which to its inhabitants is only a curiosity similar to that of John Savage's unhealthy, ageing, and soma-addicted mother, who accompanied him from the Savage Reservation. Huxley's story, unlike that of Houellebecq, leaves us with the same choice facing John Savage, symbolic of an 'eclipse of reason' and – according to Max Horkheimer – symptomatic of the 'naive rejection of reason' and the glorification of 'a historically obsolete and illusory concept of culture and individuality' (Horkheimer 2003 [1947], 38–39). Following Houellebecq this is also not much of a choice today, as reflected in more recent discussions over 'the end of nature' and its implications.

### *From the end of (human) nature to the Brave New World*

Unwittingly George Church is echoing *Elementary Particles* when he describes Francis Fukuyama's concern about the relationship between human dignity and autonomy as the basis of liberal democracy's commitment to the freedom and equality of individual citizens (Fukuyama 2002). Church calls this an 'odd argument' and explains that 'enlightened democratic governments' already grant equal rights to 'those with great intelligence, strength, and good health' and those 'with severe physical and/or mental disabilities' (Church et al. 2012, epilogue). He is right, of course, but for a different reason than he imagines.

The narrative of *Elementary Particles* confirms that an ethical appeal to freedom and dignity appears shallow when confronted directly with the cruelty of nature. What this suggests is that the desire for an 'end of nature' will overwhelm the conditions needed for an 'end of history', which is the phrase used by Fukuyama to prematurely declare the triumph of liberal democracy in the nineties. As is well known, Fukuyama made the claim that even if liberal democracy is not about to become universal or stable, there is 'a universal evolution in the direction of liberal democracy' (Fukuyama 1992, 71). The 'odd' argument he made subsequently is derived from his acknowledgement that 'there can be no end of history without an end of modern natural science and technology' (Fukuyama 2002, 7).

He is hardly alone in the view that biotechnology threatens to undermine 'human nature as a meaningful concept [that] has provided a stable continuity to our experience as a species' (ibid.). Many worry about the future of moral autonomy of the individual, as the highest human good in liberal societies, a concern shared by other notable liberals like Jurgen Habermas, who sees the identity and self-understanding of the human species as 'the context in which our conceptions of law and morality are embedded', crucial for our capacity to see ourselves as the authors of our own life histories (Habermas 2003, 23–25).

Such arguments must be particularly 'odd' to Church, who has lived through the experience of how the life sciences began demonstrating the enormous overlap between genotypes, whether of humans or any other species. More recently, by now synthetic biology has gone beyond comparing maps of genes and genomes and even beyond the transfers of genetic materials between two species (horizontal gene transfer). There are now redesigned gene expressions, metabolic pathways, and complex biological systems that together include 'parts of hundreds of previously separate species'. As Church explains, this is a 'massive and intentional exchange' that shows how 'the interspecies barrier is falling as fast as the Berlin Wall did' (Church et al. 2012, epilogue, #7).

Considering these types of developments, what realistic chance is there for Fukuyama's appeal? Fukuyama expects the state to be able to guarantee the identification of man with reason by means of protections for the unity and the continuity of human nature as it informs the classification and qualification of human rights. Not only does this imply a degree of coherency between the knowledge about genes and political philosophies about nature and human nature that is not there, but it would imply that reason is guaranteed by the state and experts. This drastically changes how reason is normally conceived in liberal theory. At least since Kant (with the categorical imperative), reason is conceived as being independent from, or at least a limit on, the powers of governments over their citizens rather than decided upon by those with influence and power, like experts and officials discussing the merits of imposing limits on genetic engineering.

Such an 'end to nature' logically ends up as a 'fetish split' between science and ethics, with the latter reduced to the illusion of human autonomy, reason, and dignity (Žižek 2004, 194). The ethical appeal to the protection of the human, like that of Habermas and Fukuyama, ends up staring straight into the faces of children with catastrophic illnesses, would-be mothers, or the suffering of animals that might end if only we eat synthetic meat and so on. Meanwhile the split leaves a wide open space to be occupied with the commodity fetishism that is characteristic of how the life sciences are governed. For example, the genetic differences between humans and other species might be important as an argument in support of human exceptionality, but the same differences can be turned into applications and represent potential value for markets in therapeutics and medicines. This is only one example, but what it shows is how a bio-ethical approach to human exceptionality would need to be sustained in the face of what Church calls 'an addiction to foreign gene products' (Church et al. 2012, epilogue, #7). Unlike societies' dependence on fossil resources, foreign gene products cannot be resisted. At most it is a habit to get under control as inevitably the (bio-)masses demand a wide range of new inventions.

The addition mentioned by Church implies a different type of escapism. The escape from one type of addiction (oil) to another implies the recognition that the humanity to be defended was never intact to begin with. First, 'we never had dignity, autonomy and reason in the first place' (Žižek 2004, 130). Second, the escape from one type of lack of humanity to another corresponds to a process wherein contemporary society is characterised as the Savage Reservation of the Brave New World. Such a state of nature is a familiar place to escape from, both in terms of how to establish a social contract in political theory and when the escape is seen in terms of the two types of biological metaphors discussed earlier. How is the tension between the biological languages of ecological limits and limitless growth (that underpin the concept of sustainable development) resolved in such a way that the promise of a return to nature can be sustained? In this regard the addiction to foreign gene products foregrounds the constant demand for new cures, new foods or types of energy, and similar types of features that give direction during the journey to a new harmonious social space. Of course, such a return to nature inevitably will turn out to lead to a space that is not nearly as harmonious in its break with the 'old' society. After all, such a new 'world in common' will already have been comprehensively settled, as it is already being inhabited by a variety of 'species of biocapital' with features that demonstrate continuity with the relation to the natural world that we are escaping from (Helmreich 2008).

Such a critique of the alternative is not intended to simply resort to another challenge of the determinisms that characterise much of the world of science and technology. The remainder of the chapter seeks to open up and rethink the alternative, engaging with politics of imagination that challenge what counts as the alternative. We have yet to follow the return to nature (from its shape in synthetic biology and the end to human exceptionalism of Church and Houellebecq) to the recent attempts to draw on the language of access, openness, and freedom to transform the life sciences.

## Biohacking and open source seeds

### *Re-contextualising the biohacker*

> They believe that information has almost mystical power of free flow and self-replication, as water seeks its own level or sparks fly upwards and lacking any moral code, they confuse inevitability with Right. It is their own view that one day instead of Feeds terminating in matter compilers, we will have Seeds that, sown on the earth, will sprout up into houses, hamburgers, spaceships, and books – that the Seed will develop inevitably from the Feed, and that upon it will be founded a more highly evolved society. [. . .] when he spoke again, it was in a clearer and stronger voice. Of course, it can't be allowed.
>
> Neal Stephenson, *The Diamond Age* (1995, 384)

The quote above, from Stephenson's *The Diamond Age*, imagines a geopolitical conflict between societies and social identities based on a system that can assemble molecules in any complicated structure. It is called 'the Feed', while others are adopting a rivalling system, called 'Seed Technology' that seeks to restore agriculture.

The two technologies are familiar. 'The Feed' resembles how DNA synthesis (see Chapter 6) and 3D printers, like 'the Seed', are seen as technologies that will radically transform design, production, distribution, and consumption. The novel describes the Seed as the next challenger, seeking to undermine the control over how information is being re-materialised with the Feed. The 'age of the Seed' threatens to displace the social order organised around this control by a new version of the idealisation of free flow and self-replication. Once more the resistance associates itself with the dictum of the hacker movement that 'information wants to be free' (see Moglen 2003).

The story begins with two children using the menu for freely printed items from a booth that is able to materialise objects by realigning their molecules. The children use it to get some flasks of water, chopsticks, nano-surimi, and isolation-blankets for the night. They have run away, and through their eyes we see a world wherein few people behave in accordance with the protocols that guarantee the functionality, safety, and branding of such objects. Of course breaking protocol is illegal, but the reality is now that many towns are continuously in a smog-like fog of nano-devices that swirl around in the food, water, and air as well as in human bodies – their lungs, blood, and so forth.

Such problems with 'waste' show the dark side of concepts like the circular economy and cascade models discussed earlier. These seek to integrate sustainability and development within the rhetoric of the global bio-economy and presume cheap basic resources. What, however, will happen when materials have become cheap and as a result are abandoned more easily? They are made from millions of microscopic particles that are often dysfunctional, do not break down according to protocol, or were deliberately released. Some of these can even be hacked, for instance to exchange data, perform computations, run programmes, or gather intelligence without using sanctioned matter-compilers.

What the story engages with is the fears attached to how new technology inevitably becomes 'easy'. This fear is not just science fiction, as is easily demonstrated by how commonplace the subject of "biohacking" has already become. Consider how George Church observes that

> as challenging as it might be to make synthetic biology research safe and secure within an institutional framework such as a university, industrial, or government lab, matters take a turn for the worse with the prospect of 'biohackers', lone agents or groups of untrained amateurs, working clandestinely, or even openly, with biological systems that have been intentionally made easy to engineer. The problem with making biological engineering techniques easy to use is that it also makes them easy to abuse.
>
> (Church et al. 2012, epilogue)

Church establishes continuity between the figure of the biohacker and how hacking has increasingly come to refer to the opposite of the benevolent computer experts who provide a voluntary service to society. Increasingly the popular meaning of the hacker is that they are outsiders who need to be held responsible for the unauthorised downloading of content, the leaking of classified information, or as the programmers of software who have the express intention to break the law of the land, whether intellectual property laws, national security, or otherwise (see Söderberg 2010).

The two types of hacking are not a choice but a false dilemma. It assumes a choice between experimentation as the prerogative of responsible science and innovation of established synthetic biology whose work is safe and secure or experimentation as the source of excess when involving irresponsible and malicious biohackers who work in isolation or outside of the scientific community. The latter implies that the biohacker turns into a living embodiment of the outside, a threat rather than a challenger who seeks to constructively transform laboratories into social spaces that need to be as inclusive and communitarian as possible as a condition for technological change to be beneficial for society. Simultaneously the biohacker turns back into a desirable object of identification in the sense that it affirms a fascination with experimentation as an object that should take place out of the reach of conventional political differences as well as the tangle of rules, regulations, and codes of conduct adhered to by the professionals.

There is no choice: the biohacker shows how difficult it has become to critically engage the life sciences constructively. The proximity of the figure of the biohacker to synthetic biology inevitably ends up undermining its challenge in the conventional antagonisms that surround genetic engineering. Let's consider this outcome in light of how the future is imagined in *The Diamond Age*. What is instructive in considering this version of the alternative is how Stephenson's narrative shows us how 'Feed technology' has made the farming of rice paddies into something superfluous.

This echoes a familiar crisis of identity, how contemporary technology plays a major part in the undermining of farming systems as based on the human hand in seed selection and crop diversity.

The novel turns to agriculture as a domain in which the distinction between nature and technology never held to begin with. What happens in the novel, however, is that certain societies are hit hardest by the new technological paradigm and have to adjust to the geopolitical consequences of how production results in a dysfunctional and fragmented biological body. Such societies no longer have the luxury of debating their preference for advanced genetic engineering (as a means to liberate biology from social constraints) or for saving nature and society from the commodification of life (by rejecting patents, corporate concentration, etc.). They have to combine both, having no option but to challenge how flows of information and materials are being controlled through 'the Feed', searching for and embracing a new relationship to the land wherein every harvest combines crops for food, fruits for medicine, and trees that produce 'synthetic rubber and pellets of clean safe fuel' (Stephenson 1995, 258).

Returning to the figure of the 'biohacker', the challenge to the laboratory can be given shape by its discussion alongside the emerging movement for 'open-source seeds'. Both draw on the rhetoric of access, openness, and freedom and might find some agreement in the diagnoses of laboratories as asocial spaces in the sense that experimentation is too often exclusively performed by researchers in white coats with multiple degrees, working with branded and patented equipment. However, open-source seeds are discussed because of how they are at once pragmatic and principled in its attempt to transform the life sciences. Rather than liberating biology from social constraints (in the Diamond Age, through biohacking, or as an escape towards the biological age), it provides a rare example of an alternative that hinges on a politics of imagination that seeks to 'renormalize what counts as biological' (Thacker 2003, 76).

### *Biohacking*

The arrival on the scene of the figure of the biohacker, in the last decade, suggests that there might be significant continuity between recent developments in the life sciences and the hacker movement. Biohackers are, like almost all of us, the inhabitants of digital environments, but more significantly, like all types of hackers, they are politically and aesthetically invested in technical practises related to information technologies. Unsurprisingly there is therefore a strong identification with the rhetoric of freedom and openness that derives from the open-source and free software movements. Yet the term biohacking is typically used interchangeably with 'do-it-yourself biologist' and 'garage biology', which is an indication that the experiment is the locus of the hackers dream of programming for its own sake – out of curiosity, playfulness, and as an exploration of the limits of the possible (see Grushkin et al. 2013; Wohlsen 2011).

Accordingly the analogy between the hacker movement and the contemporary transformation of the life sciences is not simply about the engagement with information technologies. The shared epistemological position does include coding and programming but as part of a 'remix of cultures that update a more traditional science ethos' (Delfanti 2013, 12; Penders 2011). The ideal-typical 'biohacker space' will most likely show a familiar array of computer equipment, perhaps in an old warehouse that might be demolished soon or in an inner-city basement. Surrounding old hardware will be the rudimentary tools of biology, which might be lying around, piled up, or visibly in use and under construction. Other hacker spaces, however, will be very different, with housing facilitated by prestigious institutes and with clear financial prospects. Irrespective of these differences, however, there will be microscopes, centrifuges, incubators, and spectrometers as well as the more mundane vials, refrigerators, and microwaves that might have been acquired for next to nothing. The instruments could be self-made, cheaply acquired, or obsolete in the eyes of most experts, but the key feature is that their usage takes place in laboratories designed to be social spaces that are as inclusive as possible and seek to realise the promise of lowering the required level of skills and knowledge.[6] This scenario suggests that it is not solely the irresponsibility of hackers that should be examined as a challenge to the life sciences. Quite the opposite, it is the priority given to playfulness, curiosity, and creativity that unsettles both the advocates and the critics of genetic engineering.

The case of the Glowing Plant project demonstrates this. In the summer of 2013 this project raised nearly half a million dollars on Kickstarter by promising backers who pledged more than $40 a packet of fifty to one hundred genetically altered seeds of an *Arabidopsis* plant (Calaway 2013). There is nothing particularly shocking about plants that glow in the dark, which is a genetically engineered trait that has been around since the mid-eighties as a research tool. In this regard it was helpful that *Arabidopsis* is a model organism that is widely used and studied. This includes the ability to produce a fluorescent protein that causes emission of a green-blue light, using the enzyme *Luciferase*, which is known for making fireflies and some fungi and bacteria glow (ibid.). Therefore it was not the promise of creating genetically modified plants that was controversial but the promise of its distribution at low cost and to anyone who asked.

Eventually it turned out to be complicated to create such plants as products that can be shipped to consumers, but the combination of promising a glow-in-the-dark plant with the provocative rhetoric worked. It quickly attracted around 8,000 contributors interested in having their own glowing plants wherever they wished and to be replicated by anyone with only the most basic types of laboratory equipment (ibid.). The project asked, 'What if we used trees to light our streets, instead of street lamps?' and what if '*Avatar*'s Glowing Garden Becomes a Reality?'[7] This strategy divided the community of biohackers and DIY biologists. For example, Bio-Curious, a well-known group in the US, were involved in the early stages of the project but withdrew because their code of conduct forbade the release of genetically modified organisms from their laboratories. Nonetheless there are few legal obstacles in the US, where the distribution of glowing plants is not forbidden or regulated by law, as the plants that are distributed are neither food nor micro-organisms. Moreover, the European and US codes of conduct for biohackers are usually not very specific in their appeal to principles like safety, security, access, responsibility, care for the environment, and so on.[8]

While the project was not illegal and was being marketed with some success, the controversy quickly escaped the confines of the community of biohackers and the participants in the project. Soon the Glowing Plant project had been reframed in terms of the familiar contours of the GM controversy. Once again the risks of genetic engineering were the topic of discussion rather than the merits of lowering the threshold to experimentation in improvised laboratories. The ETC Group drew attention by renaming the Kickstarter campaign Kickstopper, including the glowing plant in its campaign to regulate the release of synthetic DNA into the environment. Soon Kickstarter decided to withdraw its support for projects that reward backers with genetically modified organisms. As a result the crowdfunding–based marketing strategy could no longer rely on the principle website used by biologists and biology enthusiasts to post their projects and funding targets online.

What is significant about the project's launch is that it reveals the figure of the biohacker as a dream of an outside, imagined as a space wherein creativity flows unchecked because the normal rules do not apply. This dream might be shared with many professionals engaged in genetic engineering but only as long as biohacking refers to a few experimental spaces, amateurish, artistic, or at most as improvised labs closely associated with professionals. Here the discussion of the risks are not those prioritised by George Church, who is concerned about asocial loners and the inventions of amateur communities disconnected from responsible professionals. The potential for abuse is then about biosecurity risks, while concern over the environmental risks of plant biotechnology motivated the push back in the case of the Glowing Plant project. Therefore the perspective is not that of the professional and established synthetic biology community but of those who identify the seed as a symbol of the struggle against the neoliberal project of restructuring the social and natural worlds around the narrow logic of the market. Rather than affirming the counter-cultural identity of biohackers or a shared oppositional stance in regard to experimentation in the life sciences, the biohacker projects are interpreted as emblematic for the speculative subjectivity that drives the commodification of life.

Mistrust of biohacker politics is not limited to Kickstarter campaigns, even though this medium can be seen as illustrative of new styles of marketing and the branding of the usage of genetically modified seeds. In this sense the dream of biohacking as invention situated outside of established spaces that are heavily regulated is not seen as an alternative to the life science but as an intensification of genetic engineering and by extension the field of synthetic biology. While there are many examples that might not fall into such a characterization, most of the more or less confrontational practises of biohackers that include civil society, activists, and bioartists take place in the 'out program' of synthetic biology (Tocchetti 2012, 1). For example Christopher Kelty has commented that synthetic biology is the source of 'much of the hype and excitement around DIY Bio, Open Source science, bio-hacking and so forth' (Kelty 2010, 4). Considering the close ties between the figure of the biohacker and synthetic biology, it was always likely that controversy would follow. Perhaps it was even nearly guaranteed by the divisiveness of the genetic modification of seeds and how synthetic biology is seen as exacerbating the situation by designing new monocrops and related chemicals to be cultivated as a resource that can be transformed into energy or materials (e.g., bio-plastics). However, the point here is not that the outcome was predictable. Rather, the continuation of the status quo in the shape of the same diametrically opposed viewpoints on genetic engineering shows how comfortable either side is with a fantasy about the others as fully responsible for things that go wrong (as the cause of friction, why things are bad, as an agent of conspiracy and so on; see Žižek 1997, 210). Such paranoia is reaffirmed in the case of the Glowing Plant project, either as an example of irresponsible biohackers who intentionally let synthetic life forms go out of control or exaggerating environmentalists. The story works to establish a boundary around the work of a responsible community of synthetic biologists who were already beset on all side by social pressure. Alternatively the same sweeping and dramatic epic might as easily have these biohackers as agents of the synthetic biology establishment, opening up new frontiers for the manipulation of minds, bodies, and the environment.

Like many biotech start-ups, the Glowing Plant project ran out of money in 2017, and it never managed to insert all the six genes that were needed to grow plants that glow brightly. Simpler projects were launched, most notably one that aimed to create and distribute fragrant moss, a less complicated organism. Also this project turned out to be more challenging than expected, and when delivery had to be postponed the entire programme was ended.[9] Such failure is not surprising considering how often biotech start-ups fail, but it shows something relevant about the inflated rhetoric that accompanies many experimental set-ups' results. High expectations lead to disappointing results, and more relevantly it shows a politics of imagination wherein the obstacles and boundaries that are inevitably encountered into a basis for resuming the debate over the risks of GMOs. In this case the project was defined to provoke and marginalise dissenting understandings of the conventional boundaries between life and technology. Its sought to demonstrate that there is no need to reach out to find a basis in society for innovations in genetic engineering beyond the ability to attract funding. Such an interpretation of the biohacker as an oppositional figure is unwilling to break away from the established dynamic of insiders and outsiders, the social antagonistic relationship that synthetic biology seeks to escape. Like with GMOs, the result is that multiple values are constantly attaching themselves to new biological entities (and knowledge of such entities), making it highly likely that they take on a controversial meaning depending on the context they were introduced to (Vermeulen et al. 2012).

From the dramatic and even epic perspectives that surround synthetic biology, adversaries will appear to be everywhere. As we saw, some of its main protagonists are advocating a step beyond 'human exceptionality', and the transition that it presents to policy makers is promised to be a seamless socio-technical arrangement that spans the globe and has the life span of an industrial revolution. Ironically, however, this means that even an engaged audience that is most genuinely enthusiastic about developments in genetic engineering will end up seeming like a threat, even if in reality these are simply enthusiasts experimenting without provoking anyone. The question is therefore what to expect of the promise of liberating experimentation from the enclosed spaces of universities, governments, and companies. The attention for this project and the figure of the biohacker in general shows us that there is a desire for a transformation of the life sciences but also that such a promise seeks its realisation in direct relationship to how relatively new fields like synbio have invested in their own politics of imagination. While already limited and constrained in its ability to realise an alternative that seeks to 'open up' the life sciences, the emerging combination of synbio with the figure of the biohacker is discursively powerful as a claim on 'futures of change, openness and horizontality' that are distinguishable from those that characterised '20th century research and the related distribution of power' in the life sciences (see Delfanti 2011, 3–4).

### *Open-source seeds*

For quite a while now there has been a diverse alliance of advocacy groups, including farmers, indigenous people, and civil society advocacy groups, who have tried to slow the project of corporate 'globalisation' in agriculture. The aim is to challenge the control over plant genetic resources by corporations and governments that support their monopolisation. Success, however, requires an approach that empowers social groups and/or institutions with the mandate to sustain them and to facilitate their equitable use. There are therefore two strategic tasks: to effectively resist the project of neoliberalisation and to create space for the construction of alternatives. Both are linked and complementary as strategies, but the latter has only recently been realised in the shape of an alternative that ties the rhetoric of freedom to the usage of seeds (Kloppenburg 2010a, 2010b; Deibel and Kloppenburg 2015).

First, 'free seed' would be one of many examples in agriculture that distinguish themselves from the corporate appropriation of plant genetic resources, development of transgenic crops, and the global imposition of intellectual property rights. Whatever their many differences, agricultural producers of all types are faced with serious constraints on the free exchange of seeds. This undermines the development of new cultivars by farmers, public breeders, and small seed companies as well as community-based seed distribution or various types of farming that seek alternative agro-ecological paths. Globally this is the result of the influence of IP laws on the Convention on Biological Diversity (CBD) and the International Treaty on Plant Genetic Resources for Agriculture. These treaties function in ways that bind farmers and indigenous peoples more closely to the existing markets rather than to construct new and positive spaces for alternative action. For example, the call for farmers' rights diverts activist energies into protracted discussions about its juridical status in relation to the authorisation of patenting and the system for the management of genetic materials. This includes access and benefit sharing, as there is little to suggest that these types of mechanisms of global governance could ever turn into means of defending, much less reasserting or enlarging, farmer- or community-based sharing of seeds (see chapter 4 for a detailed discussion of the international situation).

By contrast, the 'free and open-source software' movement (FOSS) is exemplary as a response to the extension of the various forms of intellectual property to the source code of computer programming in the early eighties (copyright, software patents, etc.). Specifically Richard Stallman, the founder of the Free Software Foundation, made available the source code that he had programmed, using the newly introduced copyright to guarantee its availability. Anyone could make use of it (to study, modify, distribute, improve, etc.) on condition that the source code would remain available. Rather than renounce copyright, he began using it to prevent programmers of source code and users of software from losing the freedom to control part of their lives. By now there are thousands of projects and tens of thousands of software developers doing work on the basis of open licencing. Most exemplary is the case of Linux, which is free of charge, renowned as more stable and reliable than equivalents based on IP protection, and its millions of lines of source code can be shared, modified, and improved by anyone.

Against this backdrop a sustained effort was started in the early 2000s to think through the notion of 'open-source seeds'. The aim was to look for new ways to support those policies and initiatives that affirm that plant materials are a public good, freely available, while resisting others that are focused on plants as inventions and as exclusively owned by a few corporations and institutions (see Kipp 2005; Aoki 2008; Kloppenburg 2010a, 2010b; Kloppenburg and Deibel 2011; Deibel 2018). Crucially the analogy with informatics does not mean that seeds and software are somehow the same at the level of code. Quite the opposite; an alternative is needed as a way of anticipating markets and technologies that are transforming the usage and exchange of plant materials, which includes the expression of DNA in digital or electronic forms.

The example is therefore not about plant materials that will somehow become disembodied or decontextualised or be instantly transmissible across the globe as a digital technology. Instead an alternative is needed to how the language of 'freedom', 'openness', and 'access' is well on its way to being shaped in support of how the commodification of plants involves a wide array of exclusive rights over many different kinds of resources and types of knowledge – on genetic traits, sequences, databases, source code, and so forth. It is this fragmentation of the concepts of access and openness that requires an alternative that is given meaning and shaped at the intersection of the life sciences and informatics (Deibel 2013). It in this respect that the example of the Open Source Seed Initiative (OSSI) is exemplary for the potential of what is called a 'protected commons'. In the case of OSSI this concept refers to how plant materials are made available to those who will reciprocally share, while those who will not are excluded (see Kloppenburg 2010a, 2010b; Deibel 2018). The notion of 'free seed' as presented by OSSI aims to distinguish itself from various mechanisms that reflect how global treaties have affirmed the principle of *exclusion* – rather than *sharing* – as their constitutive basis.

The basic idea is that symmetry in flows of crop germplasm will be restored not by arranging payment for access to genetic resources but by working practically towards a reconstitution of the commons for plant materials used by breeders as well as farmers. Importantly this does not mean that peasants' landraces should be freely accessed and mined for genetic resources, as was the case under the common heritage regime (see Chapter 4). It should not be confused with the *open-access commons*, wherein it is unclear whether breeding with seeds is legal or whether farmers can replant seeds (see Chapter 4). Nor is it a feasible strategy to rely on governments to restore the public domain, which is unlikely considering the globalisation of intellectual property rules as well as industrialised agriculture. Of course states could simplify the various ways wherein working with seeds has become highly complex, but this is a different discussion. The protected commons, on the other hand, relies on open licencing, also known as copyleft, as a simple, elegant, and effective mechanism that requires no new law. In turn, the licence is the mechanism to support the collaboration of breeders and farmers who are intended to be the primary beneficiaries of the protected commons.

This is notably different than an earlier example, which was Biological Innovation for Open Source (BiOS). The initiative is interesting because it aimed to make available a package of patented plant biotechnologies. The package included a method for transferring plant genes, an alternative to the Agrobacterium technique that is owned by Monsanto. This technology is available on condition that any follow-up inventions are returned to the common pool. Anyone who builds upon the contributions of others must make available any improvements that are made to the other participants. The idea is that no one can enforce intellectual property rights against other members who have signed up on the same terms (Benkler 2006, 342; Kloppenburg 2010a, 2010b). BiOS's technologies have been used in various places, including in a joint venture with the International Rice Research Institute, and got a lot of attention in the press. However, the initial enthusiasm has not translated into the freedom to operate in plant biotechnology that was the aim. BiOS in its approach goes beyond a critique of the patent system by affirming the need for an alternative. It also restricted the potential users of the technologies that it made available by considering the users of any new seeds as passive beneficiaries of what is invented by others. In this respect, BiOS is similar to biohacking with its orientation towards opening up science, aiming to make experimental work easier for plant geneticists around the world.

This is much less the case when considering OSSI as the first example of an open-source model dedicated to maintaining fair and open access to plant genetic resources in order to ensure the availability of germplasm to farmers, gardeners, breeders, and their communities in future generations. Consider the OSSI pledge:

> You have the freedom to use these OSSI-Pledged seeds in any way you choose. In return, you pledge not to restrict others' use of these seeds or their derivatives by patents or other means, and to include this Pledge with any transfer of these seeds or their derivatives.
>
> (see http://osseeds.org/)

OSSI's website shows plant breeders and seed companies that the seeds that have been pledged can be used for further breeding without restrictions on condition that the OSSI pledge is included in 'any transfer of these seeds or their derivatives'. This is intended to ensure that the pledge remains in force for any crop variety bred from the pledged original. The pledge is simple and short, as its purpose is to remove as much of the complexity of the claims that might undermine the usage of seeds as possible.

In 2017 there were 375 varieties of seed shown on the website. These are not sold by OSSI but have been pledged by those who bred the varieties, registered them, and submitted the requisite materials (designation, information about the variety, a statement that affirms the pledge, etc.). Mostly these seeds involve plant material that was in the public domain and not legally restricted, although in a few cases there were contributions from breeders working for public institutions who had to receive permission. While it has a modest scope, the various plant varieties being offered are described as *freed seed.* What the letter 'd' adds is emphasis on how these seeds are free in comparison to the seeds previous relation to the enclosure of the plant genetic public domain and the decline of public breeding programmes. This is another reference to freedom as an alternative to the global dissemination of crop varieties that do not meet the needs of most farmers, that often cannot be legally saved, and that reinforce the expansion of unsustainable monocultures in conventional breeding and as the product of plant biotechnology.

### *A pledge as an alternative*

OSSI can be understood as another response to the decades of strengthening of the intellectual properties, but more important is that it establishes a different type of relationship to the language of access, openness, and freedom. A protected commons is given shape as a relation between life forms (crops, plant materials, etc.) and ways of living (farming, breeding, eating, etc.).

A protected commons is a requirement because of how breeders' rights, also called variety rights, have changed as a result of the authorization of patenting. For example, there used to be a straightforward norm called the 'breeding exemption', which used to forbid that a variety was protected with other kinds of intellectual property that conflicted with its guarantee of the availability of the variety for further breeding. The breeders' exemption guaranteed the accessibility of even the most competitive high-yielding conventional varieties that required great investments, which could be crossed without permission of another breeder. Hereby a wider range of genetic diversity would be incorporated into new varieties, and also it was considered useful because of discouraging monopoly breeding lines and eliminating the possibility that many different claims would apply to a new crop species with a combination of traits that are based on various other protected varieties. This is no longer the case. Around the world, a combination of plant variety rights and patents is no longer forbidden. The international Union for the Protection of New Varieties of Plant (UPOV) in UPOV91, unlike UPOV78, lifted its criteria in respect of patents and no longer guarantees that farmers can save seeds after a harvest, using these for replanting and redistribution to other farmers. The exact combinations that are possible as a consequence vary greatly across countries, but the result is often an expanded ability to restrict the saving of seeds by farmers, to enforce the payment of royalties, or otherwise impose restrictions on the usage of seeds.

It's not news that the patentability of living things has greatly altered the situation in regard of plant breeding and farming. This has complicated the situation to such an extent that OSSI-pledged seed does not rely directly on plant breeders' rights, for example by adding a clause to guarantee the breeders' exemption following the example of other types of open licencing. Such a clause would have a status that is legally speaking difficult to enforce, and even the pledge might be challenged in court, as OSSI states on its website. This legal situation was to be expected, because even the original copyleft licence (the GPL) continues to be challenged even though it is widely used and integrated into many different fields. The initial reaction to copylefting in any new field is often to declare that it is not legal. Legal controversy has accompanied GNU licencing in informatics and elsewhere. For example it was never completely settled whether open licencing of source code is legal, but this no longer matters because the users of Linux did not only turn out to be innovative but as a result made powerful allies with a strong interest in undermining the position in the software market of Microsoft. Similarly it might be argued that the terms of the OSSI pledge are not legally binding. After all, there are many conflicting types of intellectual property-related restrictions in place, whether through TRIPs, UPOV, CBD, or national laws. Everywhere conditions vary slightly, and there are conflicting laws and constant pressures on the availability of plant materials. As a result, it is never self-evident whether varieties that are not yet enclosed will remain available for further breeding or for farmers (Hughes and Deibel 2006; cf. Ghijsen 2007).

The legal dimensions might be interpreted cynically as a demonstration of how difficult it might be to establish the pledge. However, it could also be argued that this is not what matters most. The legal mechanism of OSSI exists already in many domains where similar issues exist. Why would there be a reason to expect that the release of seed is legally less feasible than the open licencing that is already widely recognised elsewhere? What is a more serious objection is that the control over the seed might be difficult to establish through legal struggle alone, and it is no easy task to establish an alternative network that includes farmers, progressive plant scientists as well as anybody else who hold relevant skills, tools, and knowledge to sustain and develop the biological and cultural diversity of plant varieties.

Many farmers and public scientists are frequently deeply embedded in existing norms and practises. While this might suggest a profound path dependency that makes radical change appear implausible, it could also be seen in terms of a potential for change. It implies that millions or even billions of farmers are trapped in a narrowing corporate seed market wherein there should be plenty of space for a few niches for public varieties and plant products that provide farmers and their crops a little of the support they need. Within this context nothing is free of charge, and neither are OSSI pledged seeds, which costs money as usual when ordered from those that do the multiplication of seeds commercially. Ordinary contracts apply in this case, which includes the sharing of benefits to whomever developed the variety and pledged it with OSSI. This does not violate the pledge as long as there are no restrictions on the usage of the seed that is passed on to breeders, farmers, or customers. The same applies to OSSI-pledged hybrid crops that do not reproduce their seeds as long as these remain free to use for any other purpose, including as a parent in breeding.

### *Seeds as an open-source movement*

Payments for open-source seeds should not be surprising when considering the analogy with other examples in other sectors. Freed seeds are not code, as was already explained, but neither are all the features of seeds so unique that they are incomparable to established fields elsewhere, whether as code, hardware, machinery, or otherwise.

For example the commercial multiplication of OSSI-pledged seeds resembles how any regular open-source programme or operating system will run on some device that is bought and sold commercially. These devices might sometimes be less expensive or less powerful, as is often the case with Linux-based desktop operating systems, and the most powerful supercomputers that overwhelmingly run on Linux. The point is that there are niches in which open source has been established that are involved directly with industries that are highly commercial without relying on restrictive practises possible because of IP protection. The same is true when a breeder pledges a hybrid variety that needs to be ordered each harvest. There is not necessarily any conflict, because there are established ways of earning an income or additional contracts, for example those that establish payment for commercial multiplication.

Similarly we can compare the orientation of OSSI to how a specialist or service company specialised in Linux might earn income based on its reputation rather than any exclusive claim over the building or maintaining of servers, supercomputers, or other devices running Linux. Income is in this case not based directly on the open licencing of Linux or other types of open-source programmes. The situation is the same when a breeder using OSSI-pledged seed will depend on an engaged community to find ways of getting direct commercial returns on a 'new' variety incorporating pledged plant materials. Valuing expertise and reputation turns the licence into a practical mechanism to reward engaged contributors to the protected commons. In this case that support would imply finding new ways to support those who seek to make a living based on the need to constantly adapt seeds to changing circumstances. It is the constant reciprocity of this process that characterises the protected commons. In this respect OSSI is exemplary, as it seeks the involvement of a wide range of practitioners as a community that can establish the licence as a working model towards supporting how the diversity of seeds requires their constant adaptation to resistance against viruses, its relation to soil, require irrigation, particular soil types, specific climatic conditions, and so on.

Obviously this is a challenging task that goes beyond the first consideration of OSSI, which is to establish a working relation with and between breeders and farmers using OSSI-pledged seeds. This step is the premise on which it becomes possible to begin scaling up the mechanism, linking it with participatory plant breeding and community-based marketing. For instance, the OSSI website mentions that it seeks food partnership, including restaurants and supermarkets, and mentions the possibilities for seed companies to benefit by showing that they care about the sustainability of the food system and potentially open up a new market with ethical consumers and even gardeners. As complicated as such scaling up might be practically, it is also simple like its pledge and basic like Linux. The latter has thrived because of the way global markets have turned software into an exclusive and expensive commodity, while hardware is now a cheap commodity that can be duplicated easily and reconstructed. Similarly, the type of agriculture OSSI can be considered as more basic, in that it starts out as a niche and that it promises a type of stability that is not available within the confines of commodified seed and working with the few crops that represent sizeable world markets and require cultivation under heavily controlled conditions.

Challenging the control over something of great value that has been commodified, like seeds, is never going to be easy. This is especially true when referring to local innovation trajectories, wherein seeds are used, which oftentimes are emblematic for the diversity of natural and cultural histories. What is unique, however, is that OSSI attempts to formulate an alternative that is based on the restoration of a baseline of reciprocity in regard of the sharing of the natural resources rather than information, knowledge, or equipment. There is already a small but growing number of open-source seed initiatives under way around the world; while each is different, the various ways wherein they present a challenge to enclosures and anticommons derived from intellectual property rights mirrors the diversity that is characteristic of open source models. Some might not succeed, while others might find sufficient support to consolidate a protected commons, beginning with being effective at site-specific problem solving. Its consequence, however, would be that something occurs of a wider significance, which is that the hierarchy between what counts as innovation begins to change as a result of having created room for creative capacity of individuals, universities, and variously sized firms as opposed to the handful of companies that have attained a dominant market position. It is at this point that we can speak of an alternative, one that is practical because of being principled in taking back control over the language of access, sharing, and openness that is otherwise becoming a feature of the ongoing estrangement of the environmental and technological commons (as shown in Chapters 4 and 6).

## Conclusion

This chapter brings together many different strands of thinking that were elaborated throughout the book, and this conclusion shows how the discussions of this chapter help elaborate further Foucault's understanding of the problem of sovereignty. This was the theoretical framework that structures both this chapter, which in the citation that started this chapter was about the need for 'a new right', which suggested there is a potential to 'bypass or get around' the problem of sovereignty (Foucault 2003, 27; see chapter 2). This chapter approached the relationship between nature and society as a comparison of different examples of alternatives, each with its own politics of imagination whether as a proposal for a global bio-economy, the figure of the biohacker, or an emerging movement for open-source seeds.

Crucially these examples are not solely presented as exceptions. Each time, the politics of imagination is different, and this could be established strictly in empirical terms, contextual, complex, and unique as examples of innovation. Instead the aim was to establish a critical relation to the theoretical framework, and to that end we need to briefly recall Foucault's problem of sovereignty. We began by tracing the problem of sovereignty to how the doctrine of right derives from monarchical power and originally focused on contracting individuals that form a social body. Over time the theory of right is redirected from its application to 'man-as-body' and 'man-as-species' and more recently towards the dissolution of a coherent conception of nature and human nature intensified today by 'tearing down the Berlin Wall between the world of objects and the world of subjects' (Haraway 1997, 270). The political theory for this historical trajectory was developed throughout the book, for example discussing the state-of-nature theories of Grotius and Hobbes (in Chapters 2, 4 and 6), Locke's labour theory (in Chapter 2), and Marx's 'species-being' commented on by Arendt (in Chapter 2). Against this backdrop it becomes possible to engage with the life sciences in naturalistic terms that do not revolve around 'reason' or 'dignity' as features that can be lost decisively as a result of how the life sciences undermine 'human exceptionality'. Rather, we begin to examine how particular conceptions of reason and dignity are threatened because they are based on specific political theories, each of which is an illustration of the problem of sovereignty.

The theory of right was not uniform at any point. An example is the move from the Hobbesian view of the state to the limitations on monarchical power in early liberal theory. Many of the theories that were highly influential in this period are rarely discussed today. Take the work of Samuel Pufendorf (1632–94), who wrote 'the first philosophically serious discussion of Hobbes'. In it he rejected Hobbes's notion of a natural right of life and Grotius's ideas about property in a state of nature (Tuck 1999, 156). While not often discussed today, he was among the more famous political theorists of early modernity for generations because he was the principle figure to argue that natural men often live in peace, as do sovereigns and citizens. In the period of early colonisation he advocated that the right to life and self-preservation does not imply a right to war, as most natural philosophers accepted. Rather, all property is contractual and the result of individuals having claims on each other (Pufendorf 1991; Krieger 1965; Tuck 1999; see Tully 1991). Today it is Rousseau's natural philosophy that is erroneously seen as the origin of the view of man as a social being (the noble savage), even though he argued that natural men are isolated and alone. The archetype has been carried over directly to the interpretation of 'human nature' in a technical world (John Savage or the 'invalid' Vincent Freeman in *Gattaca*) as well as to the violent monsters created through technology (monster of Frankenstein, Godzilla, etc.).

Pufendorf, on the other hand, was already aware of the close relation between the insecurity of nature and the colonial ambitions of sovereign individuals and sovereign societies, which is most often discussed in the context of the English school of international relations that studies the international system in terms of the expansion of the relations between 'civilized' societies. Taking seriously Pufendorf's relation to the problem of sovereignty implies that there is little that is 'new' about a 'new right' that Foucault proposed. The opposite of Hobbes's Leviathan can be found already at that very moment within the doctrine of right and can be followed along with the historical arc of the problem of sovereignty. First, it is quite obvious Pufendorf's contribution was marginalised: his natural philosophy challenged the need for a clear distinction between a state of nature and a state of society. That means there is no basis for the origin myth of the liberal agent, which is to say for the sovereign state and sovereign individual with rights of war and property. As James Tully explains, if it is the duty of every man 'to cultivate and preserve sociality' then there are also no perfectly isolated principles of reason and no mythical contract uniting the sovereign with society (Tully 1991, xxvi). Second, the arc of our argument about the problem of sovereignty can be followed further, bringing us directly to the Foucauldian idea of power: how 'multiple relations of power traverse, characterize, and constitute the social body' (Foucault 2003, 48, see chapter 2).

The point is therefore not to defend the entirety of Pufendorf's idea about the social contract. In his view, it was founded on the 'moral compulsion on men to unify further their convergent wills towards mutual safety, and (. . .) submit their wills to definition and execution by a single authority' (Krieger 1965, 125, see also Oestreich 2002). This has strong moralist and even absolutist overtones from a contemporary standpoint if read literally as a model for how a state should function. As a theory of right, however, it seeks to explain what unifies the social body and is exceptional among theories of right that 'basically knew only the individual and society' (Foucault 2003, 245). What matters here is that Pufendorf's theory can be interpreted along with the more familiar emphasis in Foucault's work on interventions in the life of populations, from birth to reproduction and death. The problem of sovereignty is sufficiently heterogeneous in its historical origin to rethink the ways wherein bio-objects are being reconfigured in distinctly early-modern terms, for example as a return to nature or as mandated by a series of social contracts –and, of course, as a domain outside of the state of society that is to be inhabited by species of biocapitalism.

So what than can we do to solve the problem of sovereignty? What does it mean to do 'precisely the opposite of what Hobbes was trying to do in Leviathan' (ibid., 28)? Step one is to take a similar position, a critique of the principle of sovereignty as realised through the constant mobilisation of knowledge and technology to mend its relation to the body politic. Step two, however, is to affirm the co-existence of a 'new right' on politically contested terrain, which is what the synthesis that follows represents in terms of the language of openness and freedom. The table 7.1 distinguishes the relationship between (im-)materiality and different types of commons (or rather 'life in common'). The X-axis contrasts life as information (virtual life) to its materiality (living matter); the Y-axis distinguishes a rhetoric of freedom that primarily refers to technology (freedom to operate) from the protected commons (free as in freedom).

In the top-left quadrant is the BioBricks Foundation. This example from synthetic biology prioritises 'life as information' in its approach to experimentation. DNA synthesis relies heavily on open access to data in the life sciences and its integration with more established types of biological experimentation. The upper-right top quadrant is where 'biohacking' and 'BiOS' reside. They combine an emphasis on technology that is readily available and easy to use with genetic engineering. OSSI is in the quadrant on the low right and applies a principled approach, combining an emphasis on freedom with working with seeds as the material embodiment of farming and agriculture.

Most significantly, however, the low-left quadrant is, in effect, empty. There are currently no significant examples of combining a protected commons like that of OSSI with an interface that connects to the life sciences directly. In this box there should be equivalents to the attempt to remove restrictions from the usage of source code, extending it to the removal of the restrictions on the usage of DNA as information and its re-materialisation as the products that are the result of the

*Table 7.1* Virtual life/living matter

| | *Virtual Life* | *Living Matter* |
|---|---|---|
| Freedom to operate | The BioBricks Foundation | Biohacking<br>BiOS |
| Free as in Freedom | *Recoding Synthetic Life©* | OSSI |

new chemistry. The empty quadrant on purpose has the title of Chapter 6. This chapter sought to pursue what an open-source alternative might mean within the confines of synbio. The table takes a step further, suggesting an alternative that takes the analogy with informatics to a next level, integrating 'openness' in the life sciences (left top) with emphasis on materiality in the right side of the table. Can we imagine a protected commons approach that draws on both categories?

Perhaps this appears extremely unlikely from the vantage point of their political differences today, as it presumes that those involved will at long last find ways of acting in common. From a more theoretical vantage point, however, taking the idea of the 'protected commons' provides a means of being concrete about Foucault's 'new right'. As just mentioned, this is not so much 'new' in the sense of doing something that does not exist; instead it refers to taking as a starting point the differences from within the doctrine of right in how to affirm the mythical contract uniting the sovereign with society and, today, the natural world. This contract includes the suggestion that it is only a fully sovereign individual that can make use of reason and dignity, which is a conception of freedom that mirrors the self-image of those who had 'the wealth, power, and leisure to conceptualize themselves as autonomous beings'. Rather, the notion of a 'new right' challenges the underlying myths as they are performed today with the 'exhilarating prospect of getting out of some of the old boxes' that limit our thinking about what being human means (Hayles 1999, 285–288). The idea of a 'protected commons' ties the theory and the analysis together. Unlike technologically deterministic ideas that hinge on the unity of time (such as the naturalistic and historical rhetoric of the global bio-economy), this concept extends the politics of imagination that derives from the basic logic of the OSSI pledge.

The theoretical perspective is obviously no longer solely about plant materials but shifts focus to the relation between life forms as bio-objects and ways of living as illustrations of the biopolitical.

In this context the availability of resources and types of knowledge in general can be considered in terms of a protected commons. Each possible example should be accessible on condition of the absence of restrictions. Such an extension of the protected commons is quite easy to imagine because it simply mirrors the extension of intellectual properties to new domains, intensifying the already established logic of the free and open-source movements as well as commons theory.

This applies not only to the life sciences, whether synbio, nanotech or otherwise, but also to many examples that would more easily be seen as localist or environmentalist. For example, there are now regional products and traditional knowledge that is protected. Why not make pledge products that somehow involve local or regional appreciation of the taste of crops or their colour? And similarly there are countless agronomic qualities and other aspects to local, regional or traditional varieties, crop landraces, and native breeds that can be identified. There are numerous examples, like traditional methods for breeding, food processing, tillage, conservation, and storage as well as the design of farmsteads, animal husbandry, land use systems, and the like. Rather than claiming them, they would be shared with those who similarly appreciate the relation between the product and its place of origin or innovations that are tied to local and cultural practises. Furthermore the model would work in the opposite direction as well. The creation of new and positive spaces for alternative action that are committed to a renewal of the public domain could include the release, access, and sharing of information in the life sciences provided this is no longer optional in regard of how it has been extracted from living materials and how it is being rematerialized as food, fuel, medicine, and so on. This is as simple as the OSSI pledge, which is basically contract law being reinterpreted in one of the many domains that might be turned into vehicles whereby to critically engage with experimentation in the life sciences in its relation to agriculture and the food system. It can easily include genetic engineering, because whether or not it should be the solution to a practical problem depends on a baseline of reciprocity. Its development would be conditional on a 'non-proprietary interface' freed from the restrictions that characterise the language of access and openness almost everywhere.

Such examples merely aim to show that there is no limit on the number of inventions that could be made available under a copyleft in alliance with a range of social movements and in support of a greater variety of causes. What is important is that in such a context there is little space for the naturalistic myths that promise seamless and effortless combination of growth with limits to growth, such as those that characterise the global bio-economy and sustainable development in general. Technically one could argue that languages of naturalism would continue, and therefore a type of bio-constitutionalism remains at work in such an alternative. This is true, but as a naturalism that is fully invested in the many shapes given to the escape from the crisis of the present, each of which attempt to vocally re-assert political traditions of creativity, free exchange, and reciprocity as a last chance to populate the entanglement of natures and societies of the future with engaged interpreters, histories, networks, forums, agoras, parliaments, and other instruments needed 'to compose a common world progressively' (Latour 1993). Such multiplicity, however, does not simply find its feet, become a movement, or have the many different types of activities take on the intended meaning. The puzzle has many pieces, some of which require us to 'recode life in common' as biohackers should, while others show how urgent it is to free seeds as the biological- and organic-minded crowd should, and even – why not? – one or two that show attempts to rethink the meaning of critique, as (bio-)sociologists should.

## Notes

1 The image can be found on: www.dsm.com/content/dam/dsm/cworld/en_US/documents/backgrounder-bio-based-economy.pdf (accessed on 1.03.2016).

2 For example, the image can be found here www.dsm.com/content/dam/dsm/cworld/en_US/documents/bio-based-business-seminar-from-a-technology-point-of-view.pdf (accessed on 22.01.2018).
3 See www.greenbiologics.com/index.php, www.corbion,com, www.bio-amber.com (accessed on March 2016).
4 See www.gevo.com, www.dna.com and www.avantium.com (accessed on 01.03.2016).
5 See http://cmsdata.iucn.org/downloads/manifesto_bbe_english_02.pdf (accessed on 01.03.2016).
6 Actual biohacker spaces in Europe are being referred to: the warehouse is the former location of the hack-lab of La Pailasse in the outskirts of Paris (see www.lapaillasse.org/, last checked December 2016), the basement refers to the 'biologigaragen' in Copenhagen (see http://biologigaragen.org/), and 'de Waag Society' (http://waag.org/en) has a highly visible location in the centre of Amsterdam in a fifteenth-century city gate, known for the anatomical theatre painted by Rembrandt as the 'The Anatomy Lesson of Dr. Nicolaes Tulp' (accessed on 8.09.2017).
7 See www.kickstarter.com/projects/antonyevans/glowing-plants-natural-lighting-with-no-electricit (accessed on 17.11.2017).
8 Reporting on the event includes: V. Postrel, The Kickstarter Culture Wars at: http://content.time.com/time/magazine/article/0,9171,2149613,00.html. See also Dan Hozowitz, Is Kickstarter Hostile To Science? At www.popsci.com/science/article/2013-08/kickstarter-anti-science. For the EU and US codes of conduct see http://diybio.org/codes/ (accessed on 01.03.2016).
9 See www.theatlantic.com/science/archive/2017/04/whatever-happened-to-the-glowing-plant-kickstarter/523551/ (accessed on 23.11.2017).

## References


Barthes, R. (1987) *Mythologies*. New York, NY: Hill & Wang.

Benkler, Y. (2006) *The Wealth of Networks*. Available at: www.benkler.org (accessed on 1.12.2016).

Birch, K. and Tyfield, D. (2013) Theorizing the Bioeconomy, Biovalue, Biocapital, Bioeconomics or . . . What? *Science Technology & Human Values*, 38(3): 299–327.

Brunori, G. (2013) Biomass, Biovalue and Sustainability: Some Thoughts on the Definition of the Bioeconomy. *EuroChoices*, 12(1): 48–52.

Calaway, E. (2013) Glowing Plants Spark Debate. *Nature*, 498(1): 5.

Calvert, J. and Joly, P. (2011) How Did the Gene Become a Chemical Compound? The Ontology of the Gene and the Patenting of DNA. *Social Science Information*, 50(2): 1–21.

Carolan, M.S. (2010) Mutability of Biotechnology Patents: From Unwieldy Products of Nature to Independent Object/s. *Theory Culture & Society*, 27(1): 110–129.

Church, G. and Regis, E. (2012) *Regenesis: How Synthetic Biology Will Reinvent Nature and Ourselves*. New York, NY: Basic Books.

Cooper, M. (2008) *Life as Surplus: Biotechnology and Capitalism in the Neoliberal Era*. Seattle, WA: University of Washington Press.

Deibel, E. (2013) Open Variety Rights. *Journal of Agrarian Change*, 13(2): 282–309.

Deibel, E. (2018) Open Sesame: Open Source and Crops. In: Girard, F. and Frison, C. (ed.). *The Commons, Plant Breeding and Agricultural: Research Challenges for Food Security and Agrobiodiversity*. New York: Routledge.

Deibel, E. and Kloppenburg, J. (2015) Open Licensing and Plant Varieties: On Creating a Protected Commons. In: Thomas, F., Bonneuil, C., and Boisvert, V. (eds.). *Le pouvoir de la biodiversité*. Paris: Editions Quae.

Delfanti, A. (2013) *Biohackers: the Politics of Open Science*. London: Pluto Press.

Dutfield, G. (2003) *Intellectual Property Rights & the Life Science Industries: A Twentieth Century History*. Aldershot: Ashgate.
Foucault, M. (2003) *Society Must Be Defended*. New York: Picador.
Franklin, S. (2007) *Dolly Mixtures: The Remaking of Genealogy*. Durham, NC: Duke University Press.
Frow, E., Ingram, D., Powell, W., Steer, D., Vogel, J., and Yearley, S. (2009) The Politics of Plants. *Food Security*, 1(1): 17–23.
Fukuyama, F. (1992) *The End of History and the Last Man*. New York, NY: Free Press.
Fukuyama, F. (2002) *Our Posthuman Future: Consequences of the Biotechnology Revolution*. New York, NY: Picador.
Ghijsen, H. (2007) Plant Breeder's Rights: A Fair and Balanced Intellectual Property Rights for Plant Varieties: Contribution to an On-going Debate. *Tailoring Biotechnologies*, 3(2): 79–98.
Goodman, D., Sorj, B., and Wilkenson, J. (1987) *From Farming to Biotechnology: A Theory of Agro-Industrial Development*. Oxford: Basil Blackwell.
Grushkin, D., Kuiken, T., and Millet, P. (2013) *Seven Myths & Realities About Do-It-Yourself Biology*. Available at: www.synbioproject.org/site/assets/files/1278/7_myths_final.pdf (accessed on 1.12.2016).
Habermas, J. (2003) *The Future of Human Nature*. Cambridge: Polity.
Haraway, D. (1997) *Modest_Witness@Second_Millennium.FemaleMan©_Meets_OncomouseTM:* Feminism *and Technoscience*. New York, NY: Routledge.
Hayles, N.K. (1999) *How We Became Posthuman: Virtual Bodies in Cybernetics, Literature, and Informatics*. Chicago: University of Chicago Press.
Helmreich, S. (2008) Species of Biocapital. *Science as Culture*, 17(4): 463–478.
Horkheimer, M. (2003) *Eclipse of Reason*. New York, NY: Continuum.
Horkheimer, M. and Adorno, T.W. (1983) *Dialectic of Enlightenment*. London: Allen Lane.
Houellebecq, M. (2000) *Elementary Particles*. New York, NY: Vintage Books.
Hughes, S. and Deibel, E. (2006) Plant Breeder's Rights: Room to Manoeuvre? *Tailoring Biotechnologies*, 2(3): 77–86.
Jasanoff, S. (ed.) (2011) *Reframing Rights: Bioconstitutionalism in the Genetic Age*. London: MIT Press.
Kelty, C.M. (2010) Outlaw, Hackers, Victorian Amateurs: Diagnosing Public Participation in the Life Sciences Today. *Journal of Science Communication*, 9(01): 1–9.
Kipp, M. (2005) Software and Seeds: Open Source Methods. *First Monday*, 10(9).
Kloppenburg, J. (2010a). Seed Sovereignty: The Promise of Open Source Biology. In: Desmarais, A., Wittman, H.K., and Wiebe, N. (eds.). *Food Sovereignty: Reconnecting Food, Nature and Community*. Black Point: Fernwood.
Kloppenburg, J. (2010b) Impeding Dispossession, Enabling Repossession: Biological Open Source and the Recovery of Seed Sovereignty. *Journal of Agrarian Change*, 10(3): 367–388.
Kloppenburg, J. and Deibel, E. (2011) La biologie 'open source' et le rétablissement de la souveraineté sur les semences. In: *La propriété intellectuelle contre la biodiversité? Géopolitique de la diversité biologique*. Geneva: Centre Europe – Tiers Monde.
Krieger, L. (1965) *The Politics of Discretion: Pufendorf and the Acceptance of Natural Law*. Chicago: University of Chicago Press.
Latour, B. (1993) *We Have Never Been Modern*. New York, NY: Harvester Wheatsheaf.
Lazzarato, M. (2014) *Signs and Machines*. Los Angeles, CA: Semiotext(e).
Lee, D. and Wilkinson, R. (eds.) (2007) *The WTO After Hong Kong: Progress in, and Prospects for, the Doha Development Agenda*. New York, NY: Routledge.

Levidow, L., Birch, K., and Papaioannou, T. (2013) Divergent Paradigms of European Agro-food Innovation: The Knowledge-Based Bio-Economy (KBBE) as an R&D Agenda. *Science, Technology & Human Values*, 38(1): 94–125.

May, C. (2000) *The Global Political Economy of Intellectual Property Rights: The New Enclosures?* London: Routledge.

Mert, A. (2012) *Governance After Nature at 'The End of History' A Discourse Theoretical Study on Sustainability Partnerships*. Vrije Universiteit Amsterdam. Available at: http://dare.ubvu.vu.nl/handle/1871/35432 (accessed on 01.12.2016).

Moglen, E. (2003) Anarchy Triumphant: Free Software and the Death of Copyright. *First Monday*, 4(8).

Mooney, P. (1979) *Seeds of the Earth: A Private or Public Resource?* Ottawa: Inter Pares.

Oestreich, G. (2002) *Storia dei diritti umani e delle liberta fondamentali: a cura di Gustavo Gozzi*. Roma: Editori Laterza.

Penders, B. (2011) Biotechnology: DIY Biology. *Nature*, 472: 167.

Pistorius, R. and van Wijk, J. (1999) *The Exploitation of Plant Genetic Information: Political Strategies in Crop Development*. Amsterdam: Print Partners Ipskamp.

Pufendorf, S. (1991) *On the Duty of Man and Citizen According to Natural Law*. Cambridge: University Press.

Sell, S. (2007) Intellectual Property and the Doha Development Round. In: Lee, D. and Serres, M. (1995) *The Natural Contract*. Ann Arbor, MI: University of Michigan Press.

Söderberg, J. (2010) Misuser Inventions and the Invention of the Misuser: Hackers, Crackers, Filesharers. *Science as Culture*, 19(2): 151–179.

Stephenson, N. (1995) *The Diamond Age*. London: Penguin Books.

Sunder Rajan, K. (2006) *Biocapital: The Constitution of Postgenomic Life*. Durham, NC: Duke University Press.

Thacker, E. (2003) What Is Biomedia? *Configurations*, 11(1): 47–79.

Tocchetti, S. (2012) DIY biologists as 'Makers' of Personal Biologies: How MAKE Magazine and Maker Faires Contribute in Constituting Biology as a Personal Technology. *Journal of Peer Production*, 2: 1–9.

Tuck, R. (1999) *The Rights of War and Peace*. Oxford: Oxford University Press.

Tully, J. (1991) Introduction. In: *On the Duty of Man and Citizen According to Natural Law by Samuel Pufendorf*. Cambridge: Cambridge University Press.

Veraart, F., van Hooff, G., Lambert, F., Lintsen, H., and Schipper, H. (2011) From Arcadia to Utopia. In: Asveld, L., van Est, R., and Stemerding, D. (eds.). *Getting to the Core of the Bio-economy: A Perspective on the Sustainable Promise of Biomass*. The Hague: Rathenau Instituut.

Vermeulen, N., Tamminen, S., and Webster, A. (2012) *Bio-objects: Life in the 21st Century*. Farnham: Ashgate.

Wilkinson, R. (eds.). *The WTO After Hong Kong*. London: Routledge.

Wohlsen, M. (2011) *Biopunk: DIY Scientists Hack the Software of Life*. New York, NY: Current.

Yoxen, E. (1984) *The Gene Business: Who Should Control Biotechnology?* New York, NY: Harper & Row.

Žižek, S. (1997) *The Sublime Object of Ideology*. London: Verso.

Žižek, S. (2004) *Organs Without Bodies: Deleuze and Consequences*. New York, NY: Routledge.

# 8 The re-articulation of biopolitical theory in an era of informatics

This book started out with observations that demonstrated how the informatic understanding of life invariably implies its management through biopolitical techniques. Such an approach implies engaging with diverse visions, claims, and social theories, which allows to some extent for a decoding of contemporary life and living. This is, however, not sufficient on its own. While this book does exactly this throughout its chapters, it holds simultaneously that many aspects of the current theorisation of our biological present should pay much closer attention to the interplay of the old and new, the dead and the alive, the structures and legitimisation of power, and ultimately notions involving the order of things, disorder, chaos, and the void of life.

Accordingly, the book has been structured in a way that combines a quick tour of modern categories of life and their conceptual, political, and material transformations with an analysis that could unpack the changes happening within the realms of the natural – the human realm as well as the animal and plant kingdoms. We aimed to take a position that enables us to engage directly with how the making of new forms of life has become less easily intelligible to theory and manageable for political power. This requires not only that different spheres of life – involving human, animal, and plant materials – be discussed across the body of current literature in the social sciences (as we do) but also that they are considered as comparable and as deeply intertwined in the dynamic of their development as examples of the biopolitical.

For example, we prioritised how some key 'new' figures of the living are being made up, invented, and imagined as we speak. Indeed, this happens in a wide variety of settings, including those of new digital languages and platforms, political infrastructures, new life forms as seen with Synthia or Cynthiya, the hacker, and programmable biomass. Together, they form an array of modern and postmodern figures, practises, and visions of the future that we claim would not be fully captured by contemporary theorisations within the biopolitical theory today, thereby falling short in their attempts to provide more clarity on our recent past and the near future. From the very start the book has discussed why current theories remain asymmetrical and partial in their treatment of biological, genetic, or synthetic life and also in their relation to politics, economics, and technique. We have frequently referenced key texts of the 'Bioeconomy School' and their theorisation about key concepts such as bio-economy, biovalue, or biocapital (e.g., Sunder Rajan 2006; Thacker 2005; Cooper 2008) as a demonstration of the modern mindset that many post-modern (including post-Marxist) theories have adopted, which demonstrate an unnecessarily restricted species-centric view (e.g., limited to a particular category of plants, animals, humans, etc.). Thus the potential for application of their argument is limited to particular cases and theoretical points of view.

This does not solely apply to the 'Biocapital School', which we appreciate for its attempt to engage with social theory; it is valid similarly for various Foucault-inspired writers today to the extent that they tend to either focus on micro-level governance practises of certain species or seek a global perspective on the living as a bewildering multiplicity. More generally, this is a predicament that does not originate with an overly narrow nexus of social theory and science and technology studies focused on biology; quite the opposite – these are attempts to conceptualise within a suffocating context wherein it is increasingly hard to suspend the notion that ontological priority should continuously and at every point in social scientific or philosophical research be assigned to the radical heterogeneity of reality. This position implies that the reliance on concepts in theoretical approaches is suspect because the empirical world cannot be grasped fully by them (see Söderberg 2017). It is beyond the scope of this book to develop the latter point, but we find (with Söderberg) that such a position is restrictive and wish to add weight to the call for opening such theoretical tensions to debate and inquiry.

To methodologically and empirically understand the rapidly shifting connections between living beings or living materials, along with the politics surrounding them, we cannot just be critical of distinctively modern ways of understanding life and politics and implications such as the lack of more general reflection on categories of life as conceptualised in the modern era. The point is not merely to establish that life forms are transforming and dissolving through technological advances and interests in mixing previously distinct categorization schemes for living, nor is it sufficient to point out how the management and governing of living in general today and in the near future (life forms and their control) are becoming more and more heterogeneous and patchy. Instead, the specifics of our position are visible in our opening up of what we called the Foucault-Marx synthesis – discussed theoretically in Chapter 2 and then in Chapter 7 – and we have presented case studies in this book to show how similar kinds of forces are at play in different power struggles over the definition, the institutional governance, and the practises and practical conditions that govern the access to the materials of life. We did this while engaging closely with the underlying heterogeneity of the current taxa that characterise the informatic content or biomaterial at hand.

It is by theorising the biopolitical that we sought to delineate the problem. While the remaining chapters demonstrated the implications of our position, our aim was to ultimately reach an understanding of a larger cultural context wherein the various topics we examined together underline the necessity of such theorization. We will not repeat the full theoretical argument or summarise it here. Instead, we will rather underscore the main subjects as illustrations of our perspective. We began by discussing how digital platforms and standardised languages are built to gather biological samples of humans for coordinated biomedical research (Chapter 3) and demonstrated how various geographical, territory-bound challenges in finding a common natural- language representation, digital codification, and legal structures need to be re-considered and efforts to address them re-worked if next-generation research is going to take place beyond nationalised territories and populations inhabiting them. The European-wide biomedical infrastructure discussed (BBMRI) is a good example of how infrastructural 'back-ends' (scientific collections and standardised platforms of representation similar to what Parry, 2004, observed earlier) are crucial in shaping our understanding of life in the life sciences at many levels. The EU's scientific bodies themselves are calling the building and coordination of these structures 'metasciences' – and in the case of the BBMRI it is these metasciences of the life sciences that act as the grounding for re-organising representations and power over living beings. This chapter therefore stands out as exemplary for the significance of species of epistemic platforms that drive numerous bioscientific innovations today: it is a global information-driven biomedia platform that shows how metacodes operate, shape and transform what counts as data, information, and knowledge in the life sciences.

In the two chapters that followed, we showed how the historical, institutional trajectory has shaped the way in which plants and animals are transformed into new political beings, 'genetic resources' in agricultural spheres, and how global biopolitics is pitted against the claims to sovereign ownership and access to these newly found genetic beings. While both are governed as national resources, there are significant differences. For example, species of plants are more frequently governed through a global language of access that allows performative boundary crossings where the conservation of nature becomes premised on new technological developments, transforming the very idea of agriculture and food at the global level. In turn, political conventions and conceptual redefinitions have reframed animals from biological beings into global agricultural informatics stock, genetic pools to be herded and defended against excessively narrow-focused optimisation of biological properties and production capacities. All of these redefinitions and recoding actions in the overarching global conventions point to the revival of the sovereign as the counter-force to homogenised, simplistic, and short-sighted references to either global economic powers or an abstract notion of the authority of states.

Certainly, the historical trajectory leading to the re-emergence of the sovereign implies tracing a global history of neo-liberalisation of nature and its commodification. Yet the global setting closely mirrors the discussion in Chapter 6, wherein access rights to synthetic biology were examined. In consequence of this close reading of access rights, 'openness' was presented as an alternative to exclusivity as a universal norm and to the dominant corporate-led biopolitical model for managing life today. In particular, the idea of Open Source, derived from collaboration models that entered computer programming in the late twentieth century, illustrate that the understanding of biology now relies fundamentally on informatics and visions centred on the capability to decode- and recode life as if it were computer software. The cases in that chapter included an analysis of how the BioBricks of synthetic biology work on some of the premises of Open Source and how the 'minimal cell' and 'minimal genome' initiatives allow us to rethink the implications of the idea of Open Source at the global level.

In Chapter 7, we analysed the emergence of the 'biohacker' in relation to the new 'synthetic organisms', situating both in the context of the framework of the 'global bio-economy', a vision for a comprehensive transition to a more sustainable world. Accordingly, one of the key goals with regard to this vision is to engineer malleable biomass that can become a sustainable and renewable source of energy, food, and materials (e.g., bio-plastics), all at the will of the designer. These discursive platforms are paving the way for a continuation of the priority given to chemical industries within the context of the large-scale biological restructuring of the realities of living, with their own spaces of meaning for life and living. This analysis also points towards the emergence of a new type of platform where hackers operate independently within whichever realm of power they choose to – at the kitchen sink, in an improvised laboratory, or in ways that engage directly with the materiality of experimentation as a necessary component of any type of alternative to the status quo of global biological realities. We took a further look at the Open Source Seed Initiative, which aims at removing the restrictions imposed on the usage of plants in agriculture, and at extending open licencing for seeds. Thereby we examined the question of whether a counter-economy is possible for agricultural seeds and how such an idea might apply to living and working with genetic materials in other formats and spheres of bio-economy too. Consider together how this demonstrates that the multiplication of 'new enclosures', based on restrictive interpretations of intellectual property, is being extended to the promise of an alternative (in the case of the global bio-economy), challenged partially (in biohacker practises) and actively resisted by discursively tying agro-ecological practises to the bioinformatic platforms.

Throughout these chapters, we have claimed that a one-sided contemporary reading of Foucault's biopolitical arguments is unable to grasp the resurrection of the sovereign as a major force shaping bioeconomies at global level. We explained how old theories of sovereignty are not obsolete but, in fact, necessary when thinking about what we call 'metacodes'. Each subject can be traced back to natural philosophy, either to Hobbes, Locke, Grotius, Rousseau, and Marx or to closely related literary figures whose work can be used heuristically. The point is not simply that they are interesting or relevant but that they are grounded in intricate debates about sovereignty as it is imagined as both natural and social. Thus, the chapters reveal the underlying bias of recent discussions and analyses of contemporary life. Borrowing from Foucault, there is a disappearing foundation, one that results in a cultural and theoretical predicament of the biopolitical theories, as they account for only part of the story about what's happening to life and its politics today.

To overcome these shortcomings, we reach back into theory in order to be able to look more closely at the forces shaping life today. Doing this implies reaching an understanding of the 'metacodes' that saturate what in Chapter 2 we called an extended nervous system – in a metaphor that captures how we have been wearing the whole of humankind as our skin already while information about us is increasingly intimate, ventures beneath our skin and inside our bodies, infiltrates our mother tongue, and saturates the world we even a few years ago still called natural (and certainly not informatics or 'Big Data').

## Metacodes of life

Throughout these pages, we have shown the various forces that we see working in conjunction in shaping the idea, materiality, and process of living today. These are the 'metacodes of life'. These metacodes form particular matrices for understanding life either through the modern paradigm of the sovereign (what we might call the **matrix of the recent past**) or through novel configurations of power and representation (the **matrix of the near future**). When considered as particular cultural epistemic platforms we moved from our theoretical position (opening up the concept of sovereignty) to a variety of historical trajectories that instruct our identification and approach of key technical and social processes and sites that can be analysed empirically. In these, we can identify various valuations of life and of living and related attributes coming to the fore. We can also see key idioms through which particular aspects of biological materialities and their bioinformatic renderings are conceptualised and brought to life, surrounded by contested politics of access and openness that are configured in relation to potential circulation and exchange. Finally, we can bring together our observations in a summary of how each subject is guided by utopian discourses that tie all of these elements together.

As a summary, Table 8.1 presents the various topics addressed by this book, each of which shows a specific side to the many-faceted space of metacodes that inhabit the physical human – world interface that seems to be scripted to absorb life into a 'new order of the real' that is composed of a different set of quasi-transcendentals (e.g., 'circulation', 'connectivity', and 'complexity') (Dillon et al. 2009). In this regard, each of these epistemic platforms represents and performs old and new modes of power and configurations of life that cut across taxa and contexts. We think that each epistemic platform forms its own metacode of life. Every time, a different configuration of actors, idioms, power dynamics, principles of access and ownership, and framing of utopias can be observed. The point is not, however, simply that heterogeneity is characteristic of each case. Rather, the book has been our attempt to point the way to a different type of analysis. Before lies an urgent task: to begin opening up social theory and the concepts used thus far, and to put them to use in ways that do the work of seeking empirical experimentation as essential, a temporary suspension of both the need for a critical position and engagement with empirical realities for their own sake with as its purpose to find again a view of an horizon from where to think across the complexity of the field of relations. This is where we conclude: recognising the analytical work that is required for an understanding of how the metacodes of life transform our sense of living, along with how these epistemic platforms reinforce each other and are going to continue mutating with their very real, material, and virtual meanings and power-effects as our emerging twenty-first century unfolds.

*Table 8.1* The metacodes of life and their epistemic platforms

| *Realms of sovereignty and biopower* | *Actors* | *Values and Meaning of life or living* | *Key idioms* | *Access and enclosure* | *Guiding utopian visions* |
|---|---|---|---|---|---|
| **1. The Matrix of the Recent Past** | | | | | |
| Global Market Platform | Corporate state, UN, WTO, & multinationals | Exclusively and privately owned, traded, and capitalised commodities | Innovation, commodity goods mass production, capitalization | Access to markets in global trade agreements | Global corporate responsibility without interference by sovereign governments. |
| →\|← | | | | | |
| Platform of National Territorial Sovereignty | National & regional governments, environmental mandates | Territorially rooted natural identity. Nationally regulated natural resources & maximisation of the value of local life | National genetic resources, species data inventories, biodiversity, natural difference | Access & benefit sharing with mutually agreed terms among treaty partners, national ownership, sovereign interpretation of agreements | National sovereignty over nature and globally enforced environmental regulations promoting global responsibility |
| **2. The Matrix of the Near Future** | | | | | |
| Biomedia platforms: translating biology to bio-informatics | Public–private research consortia | Life as a software business: standardisation of language and representation, information exchange, digital platforms | Biomedia, big data, database, algorithms, standardised biology and biomedical research, scalable digital technologies | Open access to information, research commons, research-driven enclosures. Digitalized biological life forms accessed through common platforms. | Sovereignty of research communities, legitimated through advanced bioinformatic governance of populations |
| Bioindustrial platforms: translating bio-informatics to redesigned life forms | Alliances of synthetic biology with big (petro-) chemistry | Life™ – branded biomass products, mass produced in biotech industries | Biomass as cheap and globally traded renewable resource and the basic material of the global bio-economy | Exclusively owned bioproducts, patented technologies derived from common platforms | Industrial ecological modernization: sustainable re-industrialisation of societies as dependent on a renewed corporate biopower |
| Creative platforms: biohacking | Amateurs, citizens, students of synthetic biology | Unrestricted creativity and experimentation on the values of life | Free flow, empowerment, DIY, biosecurity | Open laboratories, home laboratories | DIY maker-society, biopunk. Sovereignty of individuals. |
| Platform of resistance: Open-Source Agriculture | Plant breeders, farmers, and biological agriculture | The protection and conservation of nature as based on removal of the restrictions on the exchange of biological materials | Freed seeds, return to nature | Free as in freedom | Open-Source Food and agriculture as means to revitalise sovereign agro-food communities |

## References

Cooper, M. (2008) *Life as Surplus: Biotechnology and Capitalism in the Neoliberal Era*. Seattle, WA: University of Washington Press.

Dillon, M. and Lobo-Guerrero, L. (2009) The Biopolitical Imaginary of Species Being. *Theory, Culture & Society*, 26(1): 1–23.

Parry, B. (2004) *Trading the Genome: Investigating the Commodification of Bio-information*. New York, NY: Colombia University Press.

Söderberg, J. (2017) The Genealogy of "Empirical Post-structuralist" STS, Retold in Two Conjunctures: The Legacy of Hegel and Althusser. *Science as Culture*, 26(2): 185–208.

Sunder Rajan, K. (2006) *Biocapital: The Constitution of Postgenomic Life*. Durham, NC: Duke University Press.

Thacker, E. (2005) *The Global Genome: Biotechnology, Politics and Culture*. Cambridge, MA: MIT Press.

# Index

Note: Page numbers in bold indicate a table on the corresponding page.